HANDBOOK OF MODERN ELECTRONICS MATH

HANDBOOK OF MODERN ELECTRONICS MATH

SAM COWAN

Prentice-Hall, Inc.
Englewood Cliffs, NJ

Editora Prentice-Hall do Brasil, Ltda., *Rio de Janeiro*
Prentice-Hall International, Inc., *London*
Prentice-Hall of Australia, Pty. Ltd., *Sydney*
Prentice-Hall Canada, Inc., *Toronto*
Prentice-Hall of India Private Ltd., *New Delhi*
Prentice-Hall of Japan, Inc., *Tokyo*
Prentice-Hall of Southeast Asia Pte. Ltd., *Singapore*
Whitehall Books, Ltd., *Wellington, New Zealand*

Library of Congress Cataloging in Publication Data
Cowan, Sam
Handbook of modern electronics math.
Includes index.
1. Electronics—Mathematics. I. Title.
TK7835.C64 1982 512'.1'0246213 82-11260
ISBN 0-13-380485-2

Printed in United States of America

This book is dedicated to:
Glen Byars

WHY THIS BOOK WILL BE HELPFUL TO ANYONE IN THE ELECTRONICS PROFESSION...

This handbook presents, in a straightforward, clear manner, the application of mathematics to the field of electronics. It is written for active people in electronics, as a ready reference source of the mathematical formulas and methods used to solve electronic problems.

Since mathematics is basic to *all* electronic problem solving, the more knowledgeable you are in this subject, the more skilled and successful you will be in this field. No one can possibly remember all the formulas and mathematical methods used in electronics. That's what makes this handbook a necessity. When you need to know a formula, a method, or a particular solution, simply pick up the handbook and find the answer quickly. Consider the following examples:

A. Did you ever need to convert a voltage or power ratio into a decibel reading? It can be done quickly and easily without a calculator. Chapter 9 will show you how.

B. Did you ever need to build a particular logic function and the only chips you had on hand were 7400's (Nand Gates)? This handbook gives you the step-by-step method to solve the problem . . . and with the *minimum* number of chips.

C. Did you ever have trouble with a circuit analysis and related math problems on an F.C.C. exam? There's a wide range of problems *and* solutions all worked out for you in this handbook.

D. Did you ever have difficulty understanding how a phasing-type single sideband system worked? It can be simplified by eliminating certain mixing products, and placing the signals in quadrature (90° Shift) . . . this is all explained very clearly in Chapter 10.

E. Did you ever run across a prefix such as Gigo or Hecto and wonder what the decimal multiplier should be? Look it up in the handy reference tables to be found in Chapter 9.

The book is written so that it can be understood and used by the electronic and/or electrical technician, the engineer, the laboratory worker, and the advanced experimenter. It is full of *practical,* everyday solutions to a large variety of electronic problems. You will even find a Fourier analysis of waveforms and an explanation of the I and Q chromance signals used to produce the proper color on a T.V. set. This daily reference book will also be a great help to anyone trying to pass an F.C.C., C.E.T., or other electronics exam requiring the use of math formulas to derive the correct answer. This handbook is not simply a collection of theoretical data, but it is written as a practical, clear, one-stop reference for all people who work with electronics. The book also includes:

A. Numerous monograms that will enable you to determine inductive reactance, capacitive reactance, resonant frequencies, etc.

B. Norton's and Thevenin's theorems will show you how to simplify a complex electronic circuit in order to calculate the parameters for a single component.

C. The section on number systems shows you how to convert between decimal, binary, and hexidecimal numbers, quickly and easily.

D. Sets of equations are presented and worked out for determining resistor values, enabling you to properly bias transistor amplifiers.

E. Three methods are presented for simplifying logic circuits:
   1. N-Cube
   2. Karnaugh Map
   3. Algebraic

F. There is also a section on how to use load lines to solve both linear and nonlinear circuits.

This useful handbook covers the electronics math field from add, subtract, multiply and divide . . . to the calculus functions of derivative, integrals, and Fourier series. In each case, these functions are shown as they apply to *specific* electronic circuits and electronic problems.

No longer will you have to search through several different books to find the correct mathematical solution to an electronics problem. The solution is contained in this invaluable handbook, and it will ultimately save you a good deal of time, effort . . . and money.

Sam Cowan

CONTENTS

Why This Book Will Be Helpful vii

Chapter 1: USING NUMBERS IN ANY BASE SYSTEM 1

The Basics of the Decimal System, 1
Learning Other Base Systems, 2
How to Convert Between Systems, 5
Examples of Addition, Subtraction, Multiplication and Division, 9
Using Decimals and Fractions in Different Base Systems, 12
Various Ways to Code Numbers, 14
How to Use Complements, 17

Chapter 2: ALGEBRA: USING LETTERS TO REPRESENT NUMBERS 20

Section 1: How to Manipulate Equations, 20
Section 2: Using Determinants, 26
Section 3: Solving Sets of Equations, 33
Section 4: Understanding Graphs, 36

Section 5: Examples of Polar Coordinates Vs. Rectangular Coordinates, 40
Section 6: How to Manipulate Exponents and Radicals, 43
Section 7: Understanding Imaginary Numbers, 45
Section 8: Using Scientific Notation, 50
Section 9: Understanding Logarithms, 53

Chapter 3: HOW TO SOLVE PASSIVE ELECTRONIC CIRCUITS 60

Quick Reference List of Laws and Formulas, 60
Examples of Solutions Using Ohm's Law, 62
Examples of Solutions Using Kirchoff's Laws, 65
Examples of Circuit Solutions Using Thevenin and Norton, 69
Norton's Theorem, 72
Examples of Circuit Solutions Using Superposition, 74
How to Solve Complex A.C. Circuits, 77
How to Determine Resonance, 84

Chapter 4: UNDERSTANDING AND APPLYING BOOLEAN ALGEBRA. 87

Section 1: The Three Basic Gates, 88
Section 2: How to Generate Switching Functions from Truth Tables, 91
Section 3: Checklist of Boolean Algebra Rules, 97
Section 4: How to Simplify and Manipulate Equations, 100
Section 5: Using Maps to Simplify, 104
Section 6: Examples of Arithmetic Circuits, 115

Chapter 5: USING TRIGONOMETRY IN ELECTRONICS 123

Definitions of Functions, 123
Inverse Trig Functions, 126
Using Radians, 126
Checklist of Trig Identities, 127
How to Apply Trig to Alternating Current, 128
Complex Numbers, 133
Applications to Circuits, 134

Chapter 6: CALCULUS: PRACTICAL TIPS ON DERIVATIVES AND INTEGRALS 144

Section 1: Definition of Derivatives and Integrals, 144
Section 2: How to Use Derivatives and Integrals to Solve Circuit Problems, 153
Section 3: Quick Reference Tables of Derivatives and Integrals, 155
Section 4: Using Higher Order Derivatives and Integrals, 157
Section 5: Examples of Circuit Solutions Using Differential Equations, 159

Chapter 7: UNDERSTANDING INFINITE SERIES AND FOURIER SERIES..........164

Definition of Maclaurin and Taylor Series, 164
Quick Reference Tables, 168
Definition of the Fourier Series, 169

Chapter 8: EXAMPLES OF SOLUTIONS WITH MONOGRAMS AND GRAPHS............184

Section 1: Impedance Monogram, 184
Section 2: Coil Winding Monogram, 186
Section 3: Ohm's Law Monogram, 187
Section 4: Resonance Monogram, 189

Section 5: Parallel Resistor Graphs, 191
Section 6: Load Line Analysis, 192
Section 7: Smith Charts, 197

Chapter 9: QUICK REFERENCE TABLES 202

Section 1: Trigonometric Functions, 202
Section 2: Logarithms, 205
Section 3: Powers of Two, 211
Section 4: Decibels, 212
Section 5: Metric Conversion, 213
Section 6: Hex to Decimal Conversion, 214
Section 7: ASCII Code, 218
Section 8: E.B.C.D.I.C. Code, 219
Section 9: Binary to Decimal Conversion, 219
Section 10: Constants and Symbols, 224

CHAPTER 10: EXAMPLES OF SOLUTIONS TO SPECIAL PROBLEMS 225

Section 1: The One-Ohm Cube, 225
Section 2: A Resistor Ladder Network, 227
Section 3: S.S.B. Phasing System, 229
Section 4: I and Q Chromance Signals, 235
Section 5: Negative Feedback Amplifiers, 239
Section 6: Power Calculation in High Level A.M. Plate Modulation, 241
Section 7: Transistor Biasing, 245

HANDBOOK OF MODERN ELECTRONICS MATH

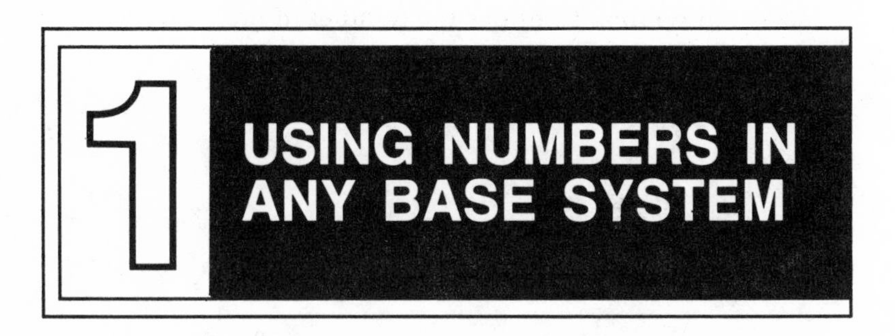

Number systems are the basics of mathematics and, therefore, the basis of understanding electronics. The fundamental operations of adding, subtracting, multiplying, and dividing must be understood in any base system. The following material will give examples of these operations, as well as other manipulations in several different base systems.

THE BASICS OF THE DECIMAL SYSTEM

The decimal or base ten system is so called because of the number of unique symbols used. By definition, there are ten symbols.

0, 1, 2, 3, 4, 5, 6, 7, 8, 9

From these symbols and the use of "positional notation," any quantity can be represented.

Example:

The number 172 is read "one hundred and seventy-two." The position of the symbols within the number is as important as the value of symbols.

In general, a number in base ten can be represented as:

$$N_1\ 10^n + N_2\ 10^{n-1} + N_3\ 10^{n-2} + \dots\dots\dots\dots$$

where N_1, N_2, N_3, ... represents one of the ten symbols, and 10^n represents the power of ten associated with the position of the symbol. Consider the number in the figure:

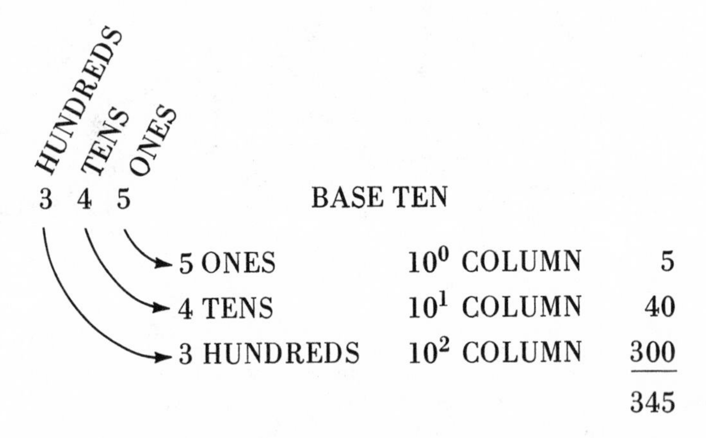

There are two basics associated with all number systems.

1. The number of symbols involved.
2. Value of the symbols' position.

The value of the symbols' position is a power of base system being used.

LEARNING OTHER BASE SYSTEMS

There is nothing special about the decimal system, except that it has the same number of symbols as humans have fingers. It is possible to use number systems with any number of digits, except zero and one.

Example 1: Counting in base five:

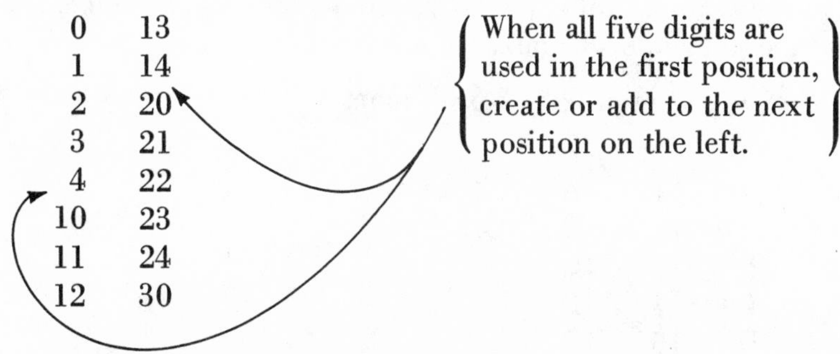

Since base five consists of five digits, when they are all used in one column, start a new column to the left. Since this is base five, the positions represent powers of five.

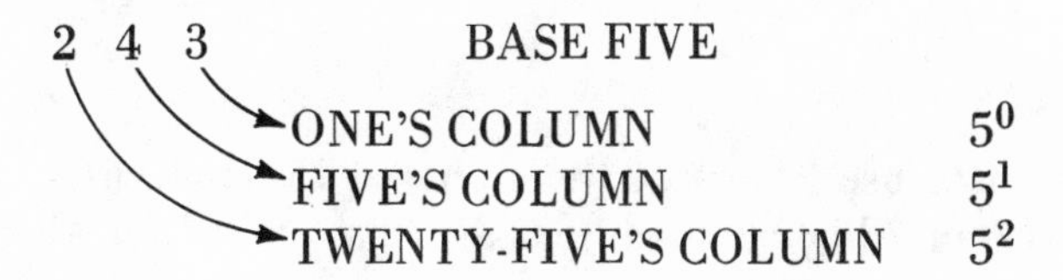

Example 2: Counting in base eight (octal):

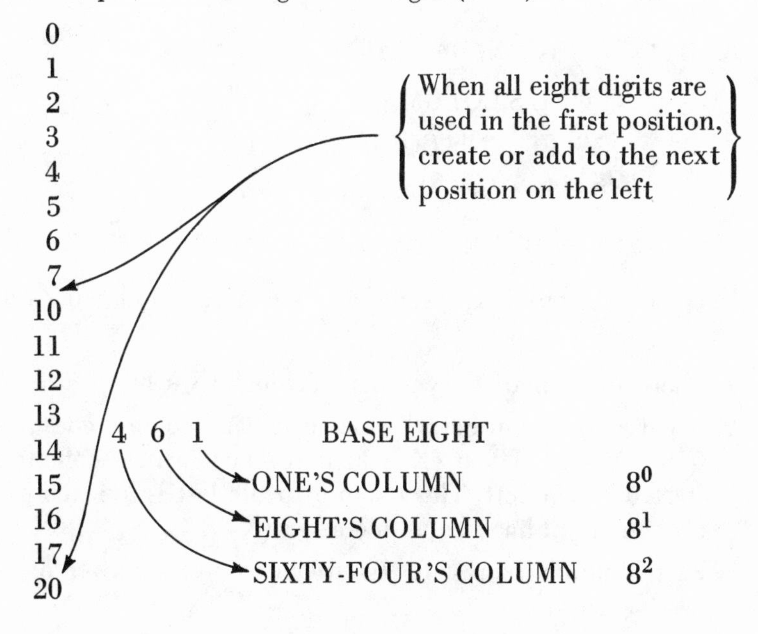

When the number of symbols required exceeds ten, it is necessary to use symbols not commonly seen in numbers. A good example is base sixteen, or the hexadecimal base.

Example 3: Counting in hexadecimal base sixteen:

0	9	11	1A	23
1	A	12	1B	24
2	B	13	1C	
3	C	14	1D	
4	D	15	1E	
5	E	16	1F	
6	F	17	20	
7	10	18	21	
8		19	22	

It is common to use letters of the alphabet to represent numbers in base systems higher than ten. In hex (short for Hexadecimal), A = 10, B = 11, etc.

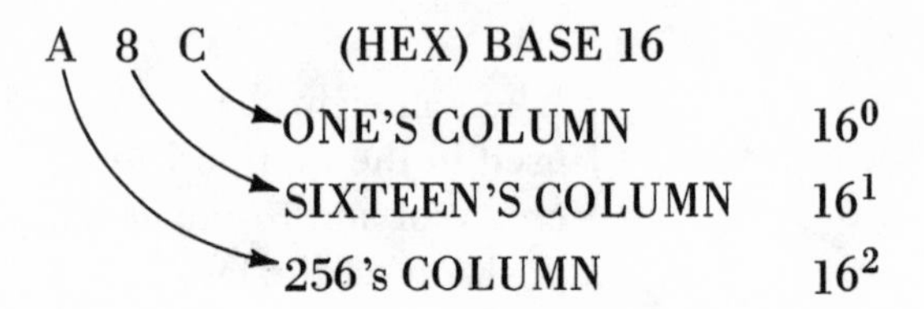

In a general number system of base X, the following rules apply:

1. There are X digits or symbols 0 through (X - 1).
2. When counting in base X, the right hand column goes from 0 to (X - 1). When (X - 1) is reached, a new column is started to the left. This column to the left is indexed each time the right hand column reaches (X - 1).
3. Each column represents (is weighted by) a power of the base system.

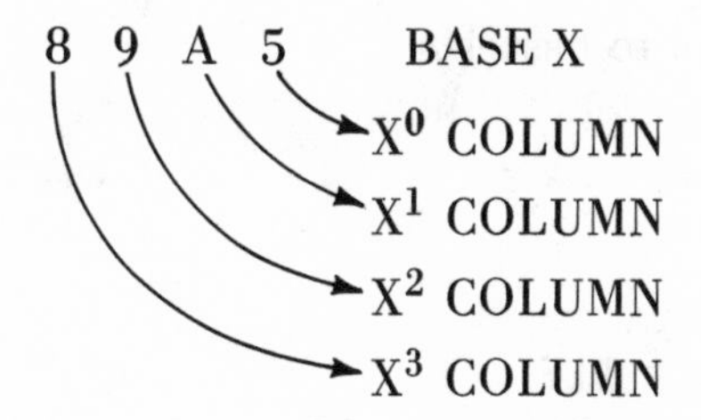

HOW TO CONVERT BETWEEN SYSTEMS

Converting between base systems reduces to two types of conversions: (1) converting from base X to base 10, and (2) converting from base 10 to base X. With these two conversions, it is possible to convert from any base to any other base. There are some simplified methods of converting between base systems that are "powers" of each other (i.e., base 2 to base 8).

Converting Base X to Base 10

Example: Convert 11010 base 2 to decimal (base 10):

16's	8's	4's	2's	1's
1	1	0	1	0

1 in the 16's column means 16
1 in the 8's column means 8
0 in the 4's column means *no* 4
1 in the 2's column means 2
0 in the 1's column means *no* 1
So, 16 + 8 + 2 = 26 base 10

Example: Convert 532 base 8 to base 10:

64's	8's	1's
5	3	2

This number has:

5	64's	5 × 64 = 320
3	8's	3 × 8 = 24
2	1's	2 × 1 = 2
		346 Base 10

Example: Convert 9A base 13 to base 10:

13's	1's
9	A

This number has:

9	13's	$9 \times 13 = 117$
A or 10	1's	$10 \times 1 = 10$
		127 Base 10

To Convert from Base X to Base 10:

1. Write the number along with the value of the column for each symbol.
2. Multiply the symbol by the value of the column.
3. Add the total.

Converting Base 10 to Base X

To convert from base 10 to base X:

1. Write the column values of base X.
2. Determine the largest value (column value X symbol) that does not exceed the base 10 number.
3. The symbol determined in step 2 is the left hand column in the base X number.
4. Subtract its value from the base 10 number.
5. With what is left from step 4, proceed with step 2, using subsequent columns to the right until the number is written.

Example: Convert 25 base 10 to base 2:

$2^5 = 32$ $2^4 = 16$ $2^3 = 8$ $2^2 = 4$ $2^1 = 2$ $2^0 = 1$

Base 2 numbers X_1 X_2 X_3 X_4 X_5 X_6

X_1 must be 0, since $32 > 25$

X_2 must be 1. This accounts for 16 of the 25 and leaves 9.

X_3 must be 1. This accounts for 8 of the 9 remaining and leaves $9 - 8 = 1$

$X_4 = 0$
$X_5 = 0$
$X_6 = 1$

The binary number is then:

1 1 0 0 1

Converting back,

16's	8's	4's	2's	1's
1	1	0	0	1

$16 + 8 + 1 = 25$ base 10.

Example: Convert 532 base 10 to base 16:

$16^3 = 4096$ $16^2 = 256$ $16^1 = 16$ $16^0 = 1$

Base 16 numbers X_1 X_2 X_3 X_4

$X_1 = 0$ $4096 > 532$
$X_2 = 2$ $532 \div 256 = 2$ plus
$2 \times 256 = 512$ $532 - 512 = 20$
$X_3 = 1$
$20 - 16 = 4$
$X_4 = 4$
Base 16 number is then: **214**

Example: Convert 341 base 10 to base 5:

Base 5 numbers $5^3 = 125$ $5^2 = 25$ $5^1 = 5$ $5^0 = 1$

X_1 X_2 X_3 X_4

$X_1 = 2$ $2 \times 125 = 250$ $341 - 250 = 91$
$X_2 = 3$ $3 \times 25 = 75$ $91 - 75 = 16$
$X_3 = 3$ $3 \times 5 = 15$ $16 - 15 = 1$
$X_4 = 1$

2331 base 5

Going through base 10, it is possible to convert between any base systems.

Example: Convert 143 base 5 to base 12:

25's	5's	1's
1	4	3

$1 \times 25 = 25$
$4 \times 5 = 20$
$3 \times 1 = 3$

$= 48$ Base 10

Base 12	$12^2 = 144$	$12^1 = 12$	$12^0 = 1$
	X_1	X_2	X_3

$X_1 = 0 \quad 144 > 48$

$X_2 = 4 \quad 4 \times 12 = 48 \quad 48 - 48 = 0$

$X_3 = 0$

40 Base 12

If the base systems are "powers" of each other, it is much easier to convert. In electronics, it is most common to deal with bases two (binary), eight (octal), and sixteen (hex).

Example: Convert 1011101 base 2 to base 8:

Since $8 = 2^3$, the number is grouped into clusters of three digits each:

```
                   421     421     421
                   ↓↓↓     ↓↓↓     ↓↓↓
(001)(011)(101) = (001)   (011)   (101)
```

Each cluster represents one octal digit.
001 = 1
011 = 3
101 = 5

Number = 135 octal

Example: Convert 11011011 base 2 to base 16:

```
           8421      8421
11011011 = (1101)    (1011)
           equals    equals
           13 or D   11 or B
```

Number = DB base 16

Example: Convert A B C base 16 to base 2:

Each base 16-digit represents four binary digits as follows:
A = ten = 1010
B = eleven = 1011
C = twelve = 1100

Binary number = (1010) (1011) (1100) = 101010111100

EXAMPLES OF ADDITION, SUBTRACTION, MULTIPLICATION AND DIVISION

Addition

When two or more numbers are to be added, they are usually shown as below, with the + sign as an operator. The result is called the sum:

```
  4391
+ 2106
------
  6497  Sum
```

In base 10 notation, if a column adds to ten or more, there must be a carry to the next column. In base X, if a column adds to X or more, there must be a carry to the next column.

Binary Addition

0 + 0 = 0
0 + 1 = 1
1 + 0 = 1
1 + 1 = 0 with a carry

```
   1        111
   11       1011
 + 11     + 1101
 ----     ------
  110      11000
```

Octal Addition

```
   1         11
   751       253
 + 054     + 265
 -----     -----
  1025       540
```

Subtraction

Subtraction is the reverse of addition. Its operator is the - sign. The top number is called the minuend, the number to be subtracted the subtrahend, and the result is called the remainder.

```
  93 minuend
- 51 subtrahend
----
  42 remainder
```

If one of the digits in the subtrahend is larger than the number above it, a "borrow" must be performed.

```
  311
  541
- 336
-----
  205
```

Binary Subtraction

```
  1011      convert        11
- 0011  →  to decimal  →  - 3
------                   ----
  1000         =            8
```

```
  1011      convert        11
-  101  →  to decimal  →  - 5
------                   ----
   110         =            6
```

When a borrow is necessary, 10 becomes 0^{10}, borrowing from the column on the left makes the column on the right a "power of two" greater.

Multiplication

$$\begin{array}{rl} 541 & \text{multiplicand} \\ \times\ 212 & \text{multiplier} \\ \hline 1082 & \\ 541 & \text{partial} \\ 1082 & \text{products} \\ \hline 114692 & \text{product} \end{array}$$

Binary Multiplication

1.

$$\begin{array}{r} 1010 \\ \times\quad 11 \\ \hline 1010 \\ 1010 \\ \hline 11110 \end{array} \quad \rightarrow \quad \begin{array}{l} \text{convert} \\ \text{to decimal} \end{array} \quad \rightarrow \quad \begin{array}{r} 10 \\ \times\ 3 \\ \hline \end{array} \qquad 11110 = 30$$

2.

$$\begin{array}{r} 111 \\ \times\ 111 \\ \hline 111 \\ 111 \\ 111 \\ \hline 110001 \end{array} \quad \rightarrow \quad \begin{array}{l} \text{convert} \\ \text{to decimal} \end{array} \quad \rightarrow \quad \begin{array}{r} 7 \\ \times\ 7 \\ \hline \end{array} \qquad 110001 = 49$$

Octal Multiplication

$$\begin{array}{r} 52 \\ \times\ 4 \\ \hline 250 \end{array} \quad \rightarrow \quad \begin{array}{l} \text{convert} \\ \text{to decimal} \end{array} \quad \rightarrow \quad \begin{array}{r} 42 \\ \times\ 4 \\ \hline 168 \end{array} \qquad 250 = 168$$

Decimal Division

```
           6.4   →  Quotient
         5)32.00 →  Dividend
        /  30
       /   --
      ↙    20
Divisor    20
           --
           00 ↘
               Remainder
```

Octal Division

```
             1   →  Quotient
Divisor    6)10  →  Dividend
              6
             --
              4  →  Remainder
```

Binary Division

```
       10                                2
101)1101    ←  convert        →       5)13
     101       to decimal               10
     ---                                --
      11                                 3
```

Octal Division

```
  36                             29
2)73    →  convert        →    2)59
  6        to decimal            4
  --                             --
  13                             19
  12                             18
  --                             --
   1                              1
```

USING DECIMALS AND FRACTIONS IN DIFFERENT BASE SYSTEMS

To understand decimals and fractions in any base system, it is necessary to state the basic rules involved in using the decimal system:

Example: Consider the fractions 1/4 and 1/3:

A. The fractions represent "one part out of four" and "one part out of three," respectively.
B. They may be converted to decimals by division.

$$\frac{1}{4} = 4\overset{.25}{\overline{)1.00}} = 0.25$$

$$\frac{1}{3} = 3\overset{.66}{\overline{)1.00}} = 0.666\ldots\ldots\ldots$$

C. When expressed as decimals, the position of a symbol indicates a power of the base (ten).

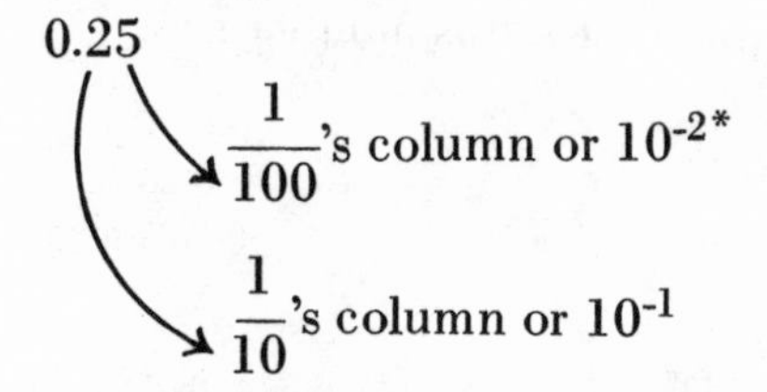

D. To multiply the two fractions:

$$\left(\frac{1}{4} \times \frac{1}{3}\right) = \frac{1}{12} \text{ or } \left(\frac{M}{N} \times \frac{a}{b}\right) = \frac{Ma}{Nb}$$

E. To divide the two fractions, invert and multiply:

$$\left(\frac{1}{3} \div \frac{1}{4}\right) = \left(\frac{1}{3} \times \frac{4}{1}\right) = \left(\frac{4}{3}\right) = \left(1\frac{1}{3}\right)$$

$$\text{or} \left(\frac{M}{N} \div \frac{a}{b}\right) = \left(\frac{M}{N} \times \frac{b}{a}\right) = \left(\frac{Mb}{Na}\right)$$

F. To add or subtract the two fractions, they must be converted to a common denominator:

$$\frac{1}{3} + \frac{1}{4} = \frac{4}{12} + \frac{3}{12} = \frac{7}{12}$$

$$\frac{1}{3} - \frac{1}{4} = \frac{4}{12} - \frac{3}{12} = \frac{1}{12}$$

*Negative powers of ten are explained in Chapter 2 Section 8.

Twelve, in this case, represents the lowest common denominator. Digits to the right of the decimal point are weighted by negative powers of the base system.

Example 1: Consider the binary number:

101.101

The position of each digit determines its weight.

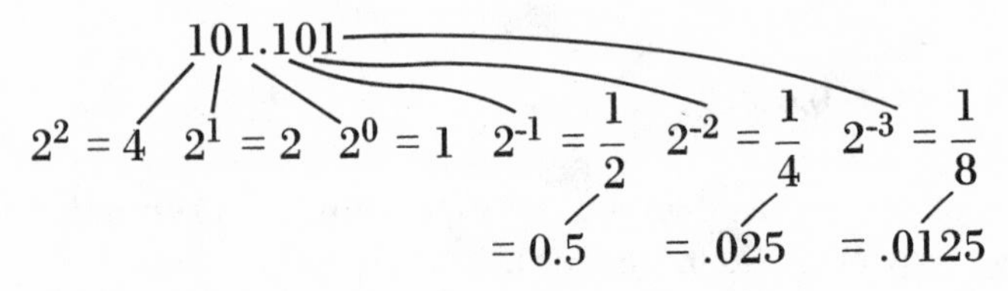

In decimal (base ten), this number is:

$$
\begin{array}{r}
4.0000 \\
+0.0000 \\
+1.0000 \\
+0.5000 \\
+0.0000 \\
+0.0125 \\
\hline
5.5125
\end{array}
$$

VARIOUS WAYS TO CODE NUMBERS

It is very common in the field of electronics to represent numbers or symbols in terms of a code. Some of the common ones follow:

B.C.D.—Binary Coded Decimal—or 8421 Code

In B.C.D., a four-digit binary number is used to represent each separate digit of the decimal number.

Examples:

8 9 3 Base 10

equals 1000 1001 0011 B.C.D.

1 5 4 Base 10

equals 0001 0101 0100 B.C.D.

Excess 3 Code

The excess 3 code is similar to the B.C.D. code, except that each decimal digit is increased by 3 before being represented as a four-digit binary number.

Examples:

8 9 3 Base 10

$8 + 3 = 11$ (1011) $9 + 3 = 12$ (1100) $3 + 3 = 6$ (0110)

1 5 4 Base 10

$1 + 3 = 4$ (0100) $5 + 3 = 8$ (1000) $4 + 3 = 7$ (0111)

Gray Code

The Gray code is a sequence of binary numbers in which one and only one digit changes in successive numbers. It is also called the "unit-distance" code.

Decimal	*Gray Code*	*Decimal*	*Gray Code*
0	0000	8	1100
1	0001	9	1101
2	0011	10	1111
3	0010	11	1110
4	0110	12	1010
5	0111	13	1011
6	0101	14	1001
7	0101	15	1000

ASCII

ASCII is the American Standard Code for information interchange. It is used to transmit information and can be thought of as a list of binary numbers representing each and every symbol. (See Figure 1-1.)

Blank Spaces Are Unassigned

	000	001	010	011	100	101		
0000	NULL	DC_0 1	ƀ	0	@	P		
0001	SOM	DC_1	!	1	A	Q		
0010	EOA	DC_2	''	2	B	R		
0011	EOM	DC_3	#	3	C	S		
0100	EOT	DC_4 STOP	$	4	D	T		
0101	WRU	ERR	%	5	E	U		
0110	RU	SYNC	&	6	F	V		
0111	BELL	LEM	'	7	G	W		
1000	FE_0	S_0	(	8	H	X		
1001	HT / SK	S_1	)	9	I	Y		
1010	LF	S_2	*	:	J	Z		
1011	V_{TAB}	S_3	+	;	K	[		
1100	FF	S_4	comma ,	<	L	\		ACK
1101	CR	S_5	—	=	M	]		2
1110	SO	S_6	★	>	N	↑		ESC
1111	SI	S_7	/	?	O	←		DEL

	Top	Side		
Example	100	1000	=	H

Figure 1-1. ASCII Code

Abbreviations

NULL	Null Idle
SOM	Start of Message
EOA	End of Address
EOM	End of Message
EOT	End of Transmission
WRU	Who Are You
RU	Are You?
BELL	Audible Bell
FE	Format Effector
HT	Horizontal Tabulation
SK	Skip
LF	Line Feed
V_{TAB}	Vertical Tabulation
DC_0	Device Control
FF	Form Feed
CR	Carriage Return
SO	Shift Out
SI	Shift In
DC_1-DC_3	Device Control
ERR	Error
SYNC	Synchronous Idle
LEM	Logical End of Media
ACK	Acknowledge
Z	Unassigned Control
ESC	Escape
DEL	Delete Idle
SO_0-SO_7	Separators

Figure 1-1. ASCII Code (contd.)

HOW TO USE COMPLEMENTS

Complements are a way of representing numbers such that subtraction can be performed by addition. Instead of subtracting a number, its complement can be added to produce the same results.

Example: The 10's complement is formed by subtracting each digit from the number 9, and then adding 1 to the least significant digit.

Normal subtraction	10's complement	
93	93	
-25	+ 75	
68	168	The 1 carry is dropped.
87	87	
-33	+ 67	
54	154	The 1 is dropped.

The 9's complement is formed by subtracting each digit from 9.

Example:

Normal Subtraction	9's complement	
59	59	
-44	+ 55	
15	114	
	+1	
	15	The carry is added to the least significant digit.

Binary Number Complements

Similar to the complements in base ten, binary numbers also have complements. Like base ten, they are used to perform subtraction by adding numbers.

The 2's Complement

The 2's complement is formed by inverting each digit in the binary number and then adding 1 to the least significant digit.

Example: Subtraction in binary is done by adding the 2's complement.

To subtract: 111011
-101011

```
Find the 2's complement:      101011
                                 ↓
                              010100
                            +      1
                            --------
                             [010101]  2's complement

Add the complement:        111011
                         +  10101
                         --------
                         1010000   The carry is dropped.
              Answer:     10000
                         --------
```

The 1's Complement

The 1's complement is formed by simply inverting each digit of the binary number.

Example: Subtraction in binary is done by adding the 1's complement, plus 1.

```
To subtract:       111011
                  -101011
                  -------

Find the 1's complement:    101011
                               ↓
                            010100

Add the complement:       111011
                        +  10100
                        --------
                         1001111

Drop the carry and add 1:   001111
                           +     1
                           -------
              Answer:        10000
```

2 ALGEBRA: USING LETTERS TO REPRESENT NUMBERS

Algebra is basic to electronics. It is the language used to express the fundamental laws governing this field. An understanding of algebra is essential to an understanding of electronics. This chapter presents all of the basic laws and methods of algebra, and shows them used for specific electronic problems.

SECTION 1: HOW TO MANIPULATE EQUATIONS

Equations are the backbone of electronic understanding and problem solving. All of the basic laws (Ohms, Kirchoff, etc.) are expressed and utilized in the form of an equation. The ability to manipulate and solve equations is essential when working in electronics.

Definition

An equation is a statement that two quantities are equal. An equality sign (=) is used to connect the two quantities. For example, consider the following equation:

$$3V + 8 = 2F + 6$$

This equation "says" that there is a quantity (V) such that, when multiplied by (3) and added to (8), is equal to the quantity (F) multiplied by (2) and added to (6). This equation can be changed and manipulated and still be the same equation, provided the changes occurred according to the rules of algebra.

Rules

(1) *The Associative Law*

$(A + B) + C = A + (B + C)$ → addition
$(AB)C = A(BC)$ → multiplication

(2) *The Commutative Law*

$A + B = B + A$ → addition
$AB = BA$ → multiplication

(3) *The Distributive Law*

$A(B + C) = AB + AC$

(4) *Multiplication*

$(+A) \times (+B) = +AB$
$(-A) \times (-B) = +AB$
$(-A) \times (+B) = -AB$
$(+A) \times (-B) = -AB$

(5) For any operation performed on one side of an equation, the same operation must be performed to the other side.

Examples:

The above rules can be used to manipulate equations and rearrange them to give desired results.

Example 1:

The equation for inductive reactance is:

$X_L = 2\pi fL$

To solve this equation for L, it is necessary to divide both sides of the equation by the quantity $2\pi f$.

$$X_L = 2\pi fL$$

$$\frac{X_L}{2\pi f} = \frac{2\pi fL}{2\pi f}$$

$$L = \frac{X_L}{2\pi f}$$

Example 2:

The equation for capacitive reactance is:

$$X_c = \frac{1}{2\pi fC}$$

To solve this equation for C:

$$X_c = \frac{1}{2\pi fC}$$

$$X_c \cdot (2\pi fC) = \frac{1}{2\pi fC} \cdot 2\pi fC$$

$$2\pi fCX_c = 1$$

Divide both sides by: $2\pi fX_c$

$$\frac{2\pi fCX_c}{2\pi fX_c} = \frac{1}{2\pi fX_c}$$

$$C = \frac{1}{2\pi fX_c}$$

Example 3:

Given: Resonance occurs when the capacitive reactance equals the inductive reactance.

Find: An equation that gives the resonant frequency when the inductance and capacitance are known.

1. $X_c = X_L$ — condition of resonance

2. $\frac{1}{2\pi fC} = 2\pi fL$ — substitution

3. $1 = 4\pi^2 f^2 LC$ — multiply both sides by $2\pi fC$

4. $f^2 = \frac{1}{4\pi^2 LC}$ — divide both sides by $4\pi^2 LC$

5. $f = \frac{1}{2\pi\sqrt{LC}}$ — take the square root of both sides

Second Degree or Quadratic Equations

A second degree equation is one in which a variable has been raised to the second power. Examples of second degree equations are:

1. $I^2 = \frac{P}{R}$

2. $X^2 + X = 4$

3. $AX^2 + BX + C = 0$

In general, a quadratic equation has two solutions, for example:

If $X^2 = 4$

$X = (+2)$ or (-2)

Since +2 squared = 4 and -2 squared = 4.

Solutions:

Simple quadratic equations can be solved by taking the square root of both sides of the equation.

$$X^2 - 9 = 0 \rightarrow X^2 = 9 \rightarrow \sqrt{X^2} = \sqrt{9} \rightarrow X = \pm 3$$

For more complex equations, i.e., those with an (x) term, this simple method of solution will not work. This type of equation requires a different solution.

Factoring:

Certain equations can be solved by factoring the equation and setting each factor equal to zero.

Example 1:

$X^2 + 5X + 6 = 0$
equals: $(X + 2)(X + 3) = 0$
since: $X + 2$
times $\underline{X + 3}$

$X^2 + 2X + 3X + 6$

equals: $X^2 + 5X + 6$

Set each factor to zero:

$(X + 2) = 0 \longrightarrow X = -2$
$(X + 3) = 0 \longrightarrow X = -3$

Example 2:

$X^2 - X - 12 = 0$
equals: $(X - 4)(X + 3) = 0$

Set each factor to zero:

$(X - 4) = 0 \longrightarrow X = 4$
$(X + 3) = 0 \longrightarrow X = -3$

This solution is quick and easy, but it applies only to equations for which factors can be found. In general, all second degree equations can be solved by substitution into the quadratic equation.

Quadratic Equation:

Given an equation: $AX^2 + BX + C = 0$

There exists a general solution which is derived as follows:

$$AX^2 + BX + C = 0$$

$$X^2 + \frac{B}{A}X + \frac{C}{A} = 0$$ Divide by A.

$$X^2 + \frac{B}{A}X = -\frac{C}{A}$$ Subtract $\frac{C}{A}$ from both sides.

$$X^2 + \frac{B}{A}X + \frac{B^2}{4A^2} = \frac{B^2}{4A^2} - \frac{C}{A}$$ Add $\frac{B^2}{4A^2}$ to both sides, this makes the left side a perfect square.

$$\left(X + \frac{B}{2A}\right)^2 = \frac{B^2}{4A^2} - \frac{C}{A}$$ Factor the left-hand side.

$$\left(X + \frac{B}{2A}\right)^2 = \frac{B^2 - 4AC}{4A^2}$$ Common denominator for the right-hand side.

$$X + \frac{B}{2A} = \pm\frac{\sqrt{B^2 - 4AC}}{2A}$$ Take the square root of both sides.

$$X = -\frac{B}{2A} \pm \frac{\sqrt{B^2 - 4AC}}{2A}$$ Subtract $\frac{B}{2A}$ from both sides.

$$X = \frac{-B \pm \sqrt{B^2 - 4AC}}{2A}$$ Combine terms.

Example 1: Solve the equation.

$$2X^2 - 3X - 4 = 0 \qquad A = 2, B = -3, C = -4$$

$$X = \frac{-B \pm \sqrt{B^2 - 4AC}}{2A}$$

$$X = \frac{3 \pm \sqrt{9 - (-32)}}{4}$$

$$X = \frac{3 \pm \sqrt{41}}{4}$$

$$X = \frac{3 + 6.403}{4}$$

$$X = \boxed{2.35075}$$

$$X = \frac{3 - 6.403}{4}$$

$$X = \boxed{-0.85075}$$

In this example, the roots (X values) turned out to be real numbers. This is not always the case, since the value under the square root sign may turn out to be negative. When this is the case, the roots (x values) will be imaginary.

Example 2:

$$3X^2 - X + 2 = 0 \qquad A = 3$$

$$X = \frac{-B \pm \sqrt{B^2 - 4AC}}{2A} \qquad B = -1, \quad C = 2$$

$$X = \frac{1 \pm \sqrt{1 - 24}}{6}$$

$$X = \frac{1 + \sqrt{-23}}{6} \qquad X = \frac{1 - \sqrt{-23}}{6}$$

Since $\sqrt{-23}$ involves imaginary numbers, the solution to this equation is as follows:

$$X = 0.1666 + 0.799J$$
$$X = 0.1666 - 0.799J$$

SECTION 2: USING DETERMINANTS

Definition

Determinants are a method by which simultaneous equations may be solved. They consist of an array of numbers as shown in Figure 2-1. When the appropriate algebraic operations are performed on the array, systems of simultaneous equations can be solved. The first part of this section shows how to manipulate a determinant and the latter parts show how determinants are used to solve equations.

$$\begin{vmatrix} X_1 & Y_1 \\ X_2 & Y_2 \end{vmatrix}$$

Figure 2-1. A Two-by-Two Determinant

Determinant Algebra

The example in Figure 2-1 is called a "two-by-two" determinant, since there are two columns and two rows. The solution to the determinant in Figure 2-1 is as follows:

$$\begin{vmatrix} X_1 & Y_1 \\ X_2 & Y_2 \end{vmatrix} = X_1 Y_2 - X_2 Y_1$$

Figure 2-2

The determinant is "solved" by multiplying X_1 and Y_2 and subtracting the product of X_2 and Y_1. In the case of "numbers" in a determinant, the solution of the determinant is simply a number as shown in Figure 2-3.

$$\begin{vmatrix} 2 & 5 \\ 1 & 3 \end{vmatrix} = [(2 \times 3) - (5 \times 1)] = (6 - 5) = \underline{\underline{1}}$$

Figure 2-3

Higher Order Determinants

The "two-by-two" determinant is easily solved by methods shown above. With a higher order determinant, this cross-multiplying method is not possible. Consider the "three-by-three" determinant shown in Figure 2-4. There are two ways to solve this determinant.

$$\begin{vmatrix} 3 & 4 & 6 \\ 2 & 3 & -2 \\ -1 & 2 & 1 \end{vmatrix}$$

Figure 2-4

1. Expansion

The first solution involves rewriting the first two columns outside the determinant and then using cross-multiplication. (See Figure 2-5.)

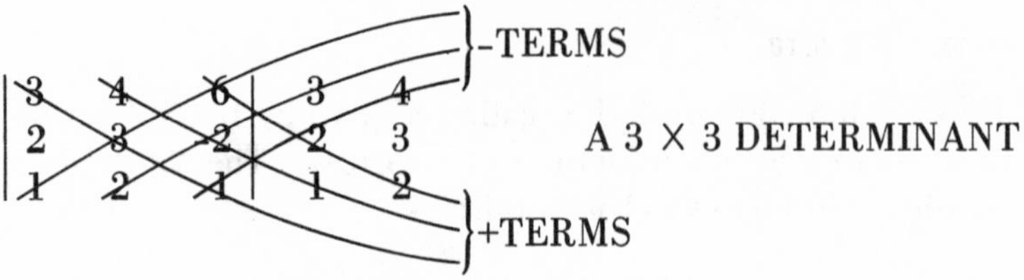

$$\begin{vmatrix} 3 & 4 & 6 \\ 2 & 3 & -2 \\ 1 & 2 & 1 \end{vmatrix} \begin{matrix} 3 & 4 \\ 2 & 3 \\ 1 & 2 \end{matrix}$$

-TERMS

A 3 X 3 DETERMINANT

+TERMS

$$(3 \times 3 \times 1) + (4 \times -2 \times 1) + (6 \times 2 \times 2) - (1 \times 3 \times 6)$$
$$- (2 \times -2 \times 3) - (1 \times 2 \times 4)$$
$$9 + (-8) + 24 - 18 - (-12) - 8$$
$$(25 - 14)$$
$$\underline{\underline{11}}$$

Figure 2-5

2. Cofactors

The second solution involves a concept called cofactors and minors. It involves reducing a determinant to a simpler form so that cross-multiplication is possible. (See Figure 2-6.)

$$\begin{vmatrix} A_1 & B_1 & C_1 \\ A_2 & B_2 & C_2 \\ A_3 & B_3 & C_3 \end{vmatrix}$$

Figure 2-6

Consider the term A_1. If we cross out the row and column associated with A_1, the remaining "two-by-two" determinant is called the minor of A_1. (See Figure 2-7.)

$$\begin{vmatrix} A_1 & B_1 & C_1 \\ A_2 & B_2 & C_2 \\ A_3 & B_3 & C_3 \end{vmatrix} \longrightarrow \begin{vmatrix} B_2 & C_2 \\ B_3 & C_3 \end{vmatrix} = \text{MINOR OF } A_1$$

Figure 2-7

In the same manner, it is possible to determine the minor of C_2. (See Figure 2-8.)

$$\begin{vmatrix} A_1 & B_1 & C_1 \\ A_2 & B_2 & C_2 \\ A_3 & B_3 & C_3 \end{vmatrix} \longrightarrow \begin{vmatrix} A_1 & B_1 \\ A_3 & B_3 \end{vmatrix} = \text{MINOR OF } C_2$$

Figure 2-8

When the method of cofactors is used to solve a determinant, it is necessary to assign a positive or negative sign to each member of the determinant. These signs are shown in Figure 2-9.

$$\begin{vmatrix} + & - & + \\ - & + & - \\ + & - & + \end{vmatrix}$$

Figure 2-9

These algebraic signs are applied to the member regardless of its actual sign. Solution of a determinant by cofactors is shown in Figure 2-10.

$$\begin{vmatrix} 1 & -3 & 2 \\ -2 & 2 & 2 \\ 4 & 1 & 3 \end{vmatrix}$$

Figure 2-10

Solution: Choose any row or column and determine the minors for each element in that row or column.

For the first column:

$$1 \times \begin{vmatrix} 2 & 2 \\ 1 & 3 \end{vmatrix} \quad -2 \times \begin{vmatrix} -3 & 2 \\ 1 & 3 \end{vmatrix} \quad 4 \times \begin{vmatrix} -3 & 2 \\ 2 & 2 \end{vmatrix}$$

Assign the appropriate algebraic sign as determined from Figure 2-9.

$$+1 \times \begin{vmatrix} 2 & 2 \\ 1 & 3 \end{vmatrix} - (-2) \times \begin{vmatrix} -3 & 2 \\ 1 & 3 \end{vmatrix} + 4 \times \begin{vmatrix} -3 & 2 \\ 2 & 2 \end{vmatrix}$$

$$1 \times (6 - 2) + 2(-9 - 2) + 4(-6 - 4)$$

$$4 - 22 - 40 = \boxed{-58}$$

These two methods can be combined to solve even higher order determinants. Figure 2-11 shows a "four-by-four" determinant and its solution.

$$\begin{vmatrix} 1 & 2 & 1 & 2 \\ 2 & 1 & 2 & 2 \\ 3 & 3 & 1 & 1 \\ 1 & 4 & 1 & 1 \end{vmatrix}$$

Figure 2-11

Expanded by cofactors:

$$+1 \times \begin{vmatrix} 1 & 2 & 2 \\ 3 & 1 & 1 \\ 4 & 1 & 1 \end{vmatrix} -2 \begin{vmatrix} 2 & 1 & 2 \\ 3 & 1 & 1 \\ 4 & 1 & 1 \end{vmatrix} +3 \begin{vmatrix} 2 & 1 & 2 \\ 1 & 2 & 2 \\ 4 & 1 & 1 \end{vmatrix} -1 \begin{vmatrix} 2 & 1 & 2 \\ 1 & 2 & 2 \\ 3 & 1 & 1 \end{vmatrix}$$

$$1 \times \begin{vmatrix} 1 & 2 & 2 \\ 3 & 1 & 1 \\ 4 & 1 & 1 \end{vmatrix} \begin{matrix} 1 & 2 \\ 3 & 1 \\ 4 & 1 \end{matrix} = 1 + 8 + 6 - 8 - 1 - 6 = 0$$

$$-2 \times \begin{vmatrix} 2 & 1 & 2 \\ 3 & 1 & 1 \\ 4 & 1 & 1 \end{vmatrix} \begin{matrix} 2 & 1 \\ 3 & 1 \\ 4 & 1 \end{matrix} = -2(2 + 4 + 6 - 8 - 2 - 3) = +2$$

$$+3 \begin{vmatrix} 2 & 1 & 2 \\ 1 & 2 & 2 \\ 4 & 1 & 1 \end{vmatrix} \begin{matrix} 2 & 1 \\ 1 & 2 \\ 4 & 1 \end{matrix} = 3(4 + 8 + 2 - 16 - 4 - 1) = -21$$

$$-1 \begin{vmatrix} 2 & 1 & 2 \\ 1 & 2 & 2 \\ 3 & 1 & 1 \end{vmatrix} \begin{matrix} 2 & 1 \\ 1 & 2 \\ 3 & 1 \end{matrix} = -1(4 + 6 + 2 - 12 - 4 - 1) = +5$$

$$0 + 2 - 21 + 5 = \boxed{-14}$$

Useful Rules:

There are several rules governing determinants that can help simplify the solution of the determinant. These rules are as follows:

Rule 1: If all of the members of any row or column are zero, the entire determinant is zero.

$$\begin{vmatrix} 0 & 0 & 0 \\ X_1 & Y_1 & Z_1 \\ X_2 & Y_2 & Z_2 \end{vmatrix} = 0$$

Rule 2: When all the members above or below the main diagonal are zero, the determinant's value is the product of the main diagonal.

$$\begin{vmatrix} X_1 & Y_1 & Z_1 \\ 0 & Y_2 & Z_2 \\ 0 & 0 & Z_3 \end{vmatrix} = (X_1 \cdot Y_2 \cdot Z_3)$$

Rule 3: When corresponding members of any two rows or columns are the same or directly proportional, the determinant is zero.

$$\begin{vmatrix} X_1 & 2X_1 & Y_1 \\ X_2 & 2X_2 & Y_2 \\ X_3 & 2X_3 & Y_3 \end{vmatrix} = 0$$

Rule 4: The *transpose* of a determinant is defined as interchanging all the rows and columns. The transpose of a determinant is equal to the determinant.

$$\begin{vmatrix} X_1 & Y_1 & Z_1 \\ X_2 & Y_2 & Z_2 \\ X_3 & Y_3 & Z_3 \end{vmatrix} = \begin{vmatrix} X_1 & X_2 & X_3 \\ Y_1 & Y_2 & Y_3 \\ Z_1 & Z_2 & Z_3 \end{vmatrix}$$

Solving Equations:

Determinants can be used to solve simultaneous equations. The method is shown by the following example:

$$AX + BY = K_1$$
$$CX + DY = K_2$$

The X solution is:

$$X = \frac{\begin{vmatrix} K_1 & B \\ K_2 & D \end{vmatrix}}{\begin{vmatrix} A & B \\ C & D \end{vmatrix}}$$

The Y solution is:

$$Y = \frac{\begin{vmatrix} A & K_1 \\ C & K_2 \end{vmatrix}}{\begin{vmatrix} A & B \\ C & D \end{vmatrix}}$$

The rules for determining the determinant solution are as follows:

Rule 1: The denominator determinant is formed by the coefficients of the unknowns, in their correct rows and columns.

Rule 2: The numerator determinant for (X) is formed by replacing the coefficients of (X) with the constants, in the correct order.

Rule 3: The numerator determinant for (Y) is formed by replacing the coefficients of (Y) with the constants, in the correct order.

Example:

Solve the following set of equations by using determinants:

$$5R + 3V = 10$$
$$2R + 2V = 3$$

The denominator determinant is: $\begin{vmatrix} 5 & 3 \\ 2 & 2 \end{vmatrix}$

The R numerator determinant is: $\begin{vmatrix} 10 & 3 \\ 3 & 2 \end{vmatrix}$

The V numerator determinant is: $\begin{vmatrix} 5 & 10 \\ 2 & 3 \end{vmatrix}$

$$R = \frac{\begin{vmatrix} 10 & 3 \\ 3 & 2 \end{vmatrix}}{\begin{vmatrix} 5 & 3 \\ 2 & 2 \end{vmatrix}} = \frac{20 - 9}{10 - 6} = \frac{11}{4} = 2.75$$

$$V = \frac{\begin{vmatrix} 5 & 10 \\ 2 & 3 \end{vmatrix}}{\begin{vmatrix} 5 & 3 \\ 2 & 2 \end{vmatrix}} = \frac{15 - 20}{10 - 6} = \frac{-5}{4} = -1.25$$

To check the results, simply substitute the results back in the original equations.

$$5(2.75) + 3(-1.25) = 10$$
$$13.75 - 3.75 = 10$$

and

$$2(2.75) + 2(-1.25) = 3$$
$$5.50 - 2.50 = 3$$

SECTION 3: SOLVING SETS OF EQUATIONS

The entire purpose of learning how to manipulate and solve equations is to apply these methods to the solution of electronic problems. Solving sets of equations is essential in most circuit analysis problems. This section gives a circuit analysis problem and then shows the different methods that can be used to solve a set of equations.

Consider the circuit in Figure 2-12. To determine the currents

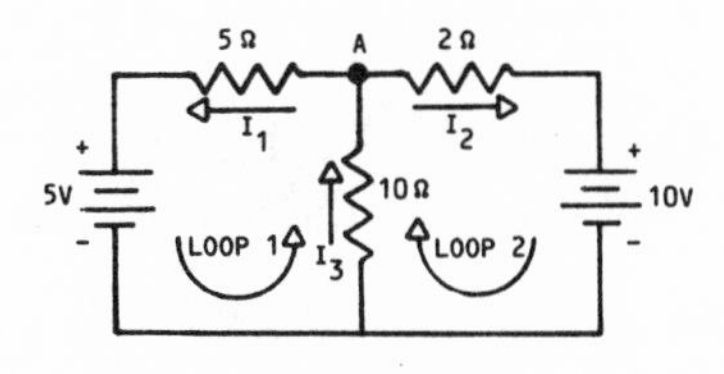

Figure 2-12

flowing in this circuit, we can write a set of simultaneous equations derived from Kirchoff's Laws. The equations can then be solved by the methods developed in the first two sections of this chapter.

A. Kirchoff's Current Law at point A.	$I_3 = I_1 + I_2$
B. Kirchoff's Voltage Law around loop 1.	$5 = 10I_3 + 5I_1$
C. Kirchoff's Voltage Law around loop 2.	$10 = 10I_3 + 2I_2$

The three equations developed by Kirchoff's Laws form three independent simultaneous equations. These equations can be solved to determine values for I_1, I_2, and I_3. For three unknowns, it is necessary to have three independent equations. In general, for (x) unknowns, it is necessary to have (x) independent equations to obtain solutions.

Solution by Determinants

The first step to solution by determinants is to arrange the equations in the form necessary. This is done below.

$$1.\ I_3 = I_1 + I_2 \longrightarrow I_1 + I_2 + I_3 = 0$$
$$2.\ 5 = 10I_3 + 5I_1 \longrightarrow 5I_1 + 0I_2 + 10I_3 = 5$$
$$3.\ 10 = 10I_3 + 2I_2 \longrightarrow 0I_1 + 2I_2 + 10I_3 = 10$$

The equations are now in the form:

$$AX_1 + BX_2 + CX_3 = K$$

The determinant solutions for I_1, I_2, and I_3 all have the same denominator, so we will solve for the denominator first.

$$\text{Denominator} = \begin{vmatrix} 1 & 1 & -1 \\ 5 & 0 & 10 \\ 0 & 2 & 10 \end{vmatrix}$$

$$\left|\begin{array}{ccc|cc} 1 & 1 & -1 & 1 & 1 \\ 5 & 0 & 10 & 5 & 0 \\ 0 & 2 & 10 & 0 & 2 \end{array}\right. = -10 - 20 - 50 = \boxed{-80}$$

$$I_1 = \frac{\begin{vmatrix} 0 & 1 & -1 \\ 5 & 0 & 10 \\ 10 & 2 & 10 \end{vmatrix}}{-80} \qquad \left|\begin{array}{ccc|cc} 0 & 1 & -1 & 0 & 1 \\ 5 & 0 & 10 & 5 & 0 \\ 10 & 2 & 10 & 10 & 2 \end{array}\right. = 100 - 10 - 50 = 40$$

$$I_1 = \frac{40}{-80} = \boxed{-0.5 \text{ Amps}}$$

$$I_2 = \frac{\begin{vmatrix} 1 & 0 & -1 \\ 5 & 5 & 10 \\ 0 & 10 & 10 \end{vmatrix}}{-80} \qquad \left|\begin{array}{ccc|cc} 1 & 0 & -1 & 1 & 0 \\ 5 & 5 & 10 & 5 & 5 \\ 0 & 10 & 10 & 0 & 10 \end{array}\right. = 50 - 50 - 100 = -100$$

$$I_2 = \frac{-100}{-80} = \boxed{1.25 \text{ Amps}}$$

$$I_3 = \frac{\begin{vmatrix} 1 & 1 & 0 \\ 5 & 0 & 5 \\ 0 & 2 & 10 \end{vmatrix}}{-80} \qquad \left|\begin{array}{ccc|cc} 1 & 1 & 0 & 1 & 1 \\ 5 & 0 & 5 & 5 & 0 \\ 0 & 2 & 10 & 0 & 2 \end{array}\right. = 0 - 10 - 50 = -60$$

$$I_3 = \frac{-60}{-80} = \boxed{0.75 \text{ Amps}}$$

A quick substitution check will show that these values satisfy the three original Kirchoff Law equations.

Solution by Algebraic Manipulation and Substitution:

Given the three equations:

$$I_3 = I_1 + I_2 \quad \text{Equation 1}$$
$$5 = 10I_3 + 5I_1 \quad \text{Equation 2}$$
$$0 = 10I_3 + 2I_2 \quad \text{Equation 3}$$

A. Solve equation 1 for I_1:	$I_1 = I_3 - I_2$
B. Substitute Step A result into Equation 2:	$5 = 10I_3 + 5(I_3 - I_2)$
C. Simplify the equation derived in Step B:	$5 = 10I_3 + 5I_3 - 5I_2$ $5 = 15I_3 - 5I_2$
D. Multiply this equation by 2:	$10 = 30I_3 - 10I_2$
E. Multiply Equation 3 by 5:	$50 = 50I_3 + 10I_2$
F. Add the equations in Steps D and E:	$60 = 80I_3 - 0$
G. Solve the equation in Step F for:	$I_3 = \frac{60}{80} = $ (0.75 Amps)
H. Substitute the value for I_3 in Equation 2:	$5 = 10(0.75) + 5I_1$ $5 = 7.5 + 5I_1$
I. Solve for I_1:	$5I_1 = -2.5$ $I_1 = \frac{-2.5}{5} = $ (-0.5 Amps)
J. Substitute I_3 and I_1 into Equation 1:	$0.75 = -0.5 + I_2$ $I_2 = $ (1.25 Amps)

These results can also be verified by substituting them into the original Kirchoff equations.

SECTION 4: UNDERSTANDING GRAPHS

Introduction

A graph is a visual representation of an equation or, more correctly, a visual representation of the relationship between two quantities. It is quite often easier to understand the relationship between quantities by viewing the graph than by looking at the equation. This section presents the basic concepts of graphs and how they are used in electronics.

Definition

Any point along the line of a graph will represent a value of

(X) and a value of (Y). The example in Figure 2-13 shows a graph of the equation $Y = X^2 - 10$, along with a table of (X) and (Y) values. The graph can represent all values of (X) and (Y) within range, while the table only represents discrete values.

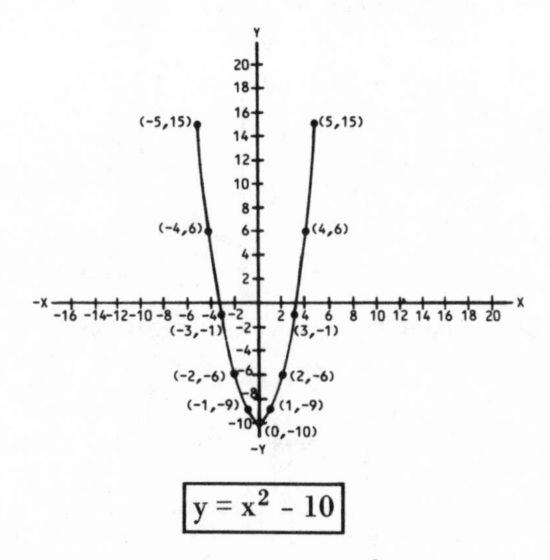

All values which satisfy the equation $y = x^2 - 10$, lie on the curve.

x	-5	-4	-3	-2	-1	0	1	2	3	4	5
y	15	+6	-1	-6	-9	-10	-9	-6	-1	6	15

Figure 2-13

Linear Equations

Of special interest in electronics is the graphing of linear equations. Several sections of this book are devoted to the solution of linear equations by various means. These equations can also be solved graphically. The graphical solution is a sort of "picture" of the solution.

A linear equation is an equation whose graph is a straight line, just as a linear component in electronics is one whose E-I graph is a straight line. In general form, a two-variable linear equation can be written as follows:

$Y = MX + B$

Where: M is the slope of the line

B is the Y intercept of the line

Any linear equation can be rearranged to look like the general form. For example:

$$3x + 4y = 5$$
$$4y = -3x + 5$$
$$y = -\frac{3}{4}x + 1.25$$
$$\left\{ \begin{array}{l} M = -\frac{3}{4} \\ B = 1.25 \end{array} \right\}$$

The graph of this equation is shown in Figure 2-14.

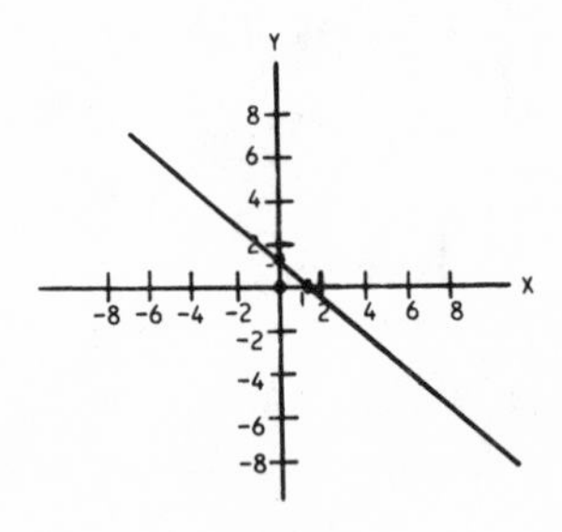

Figure 2-14

When a two-variable linear equation is put in the general form, the graph of that equation can be readily drawn by knowing the (Y) intercept (B), and the slope (M). From the two points determined by (B) and (M), the straight line graph is determined. Consider graphing the following equation.

$$2X - 4Y = 8$$

First, put the equation in the $Y = MX + B$ form.

$$2X - 4Y = 8$$
$$4Y = 2X - 8$$
$$Y = \frac{1}{2}X - 2$$
$$\left\{ \begin{array}{l} M = \frac{1}{2} \\ B = -2 \end{array} \right\}$$

To graph this equation, in Figure 2-15, the first point is on the Y-axis at Y = -2 (the Y-intercept). The second point is determined from slope as follows:

A. The slope is a positive 1/2.

B. This means that, for every unit in the +Y direction, go 2 units in the +X direction.

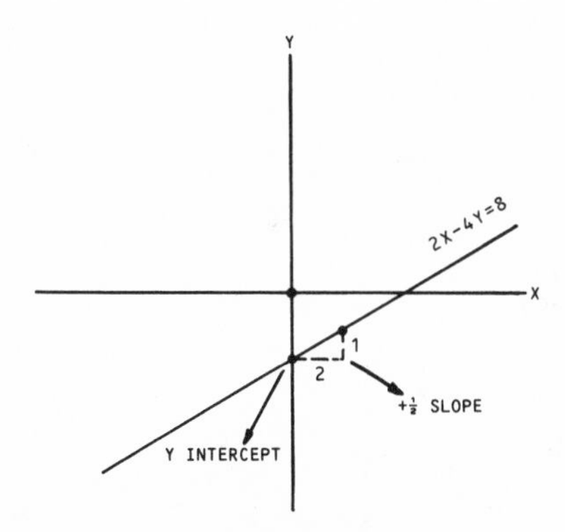

Figure 2-15

Simultaneous Equations

The solution to a set of simultaneous equations is an (X) and (Y) value that satisfies all equations involved. A graph of a linear equation is a set of (X) and (Y) values that satisfy that equation. If two linear equations are plotted on the same graph, the point at which the two lines intersect represents the solution to the two equations.

For example, consider these two equations:

$2X + 4Y = 10$ Equation 1
$4X - 2Y = 8$ Equation 2

These equations can be solved by algebraic methods:

$2X + 4Y = 10$ Equation 1
$8X - 4Y = 16$ Equation 2 times 2

$10X = 26$

$\boxed{X = 2.6}$

$$2(2.6) + 4Y = 10$$
$$4Y = 4.8$$
$$\boxed{Y = 1.2}$$

To solve these equations graphically, first put the equations in standard form:

Equation 1 $\quad Y = -\frac{1}{2}X + 2.5$

Equation 2 $\quad Y = 2X - 4$

These equations are graphed in Figure 2-16.

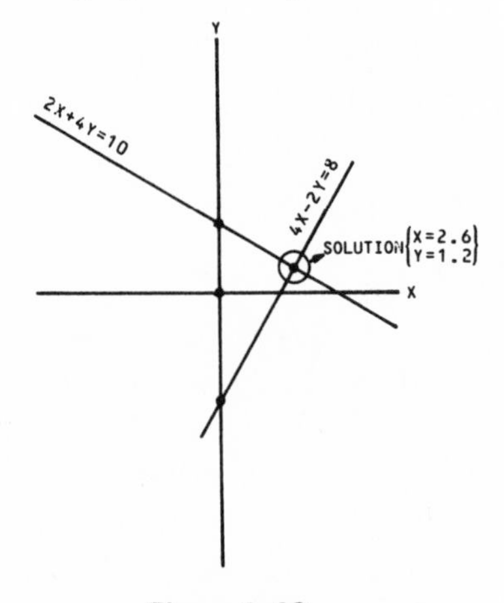

Figure 2-16

SECTION 5: EXAMPLES OF POLAR COORDINATES VS. RECTANGULAR COORDINATES

Introduction

The previous section on graphs used rectangular or "cartesian" coordinates. The two numbers that defined a point on the graph were located on perpendicular lines. Another common method of graphing or locating points is the polar notation or polar coordinates. As with rectangular coordinates, polar coordinates also require two numbers to define a point. These numbers are the magnitude and angle. Polar coordinates are the same thing as vectors.

Graphs

Figure 2-17 shows examples of points located by (X,Y) values using rectangular coordinates. Also shown are R and θ values corresponding to the distance from the center of the graph (R) and the angle from the X coordinate θ.

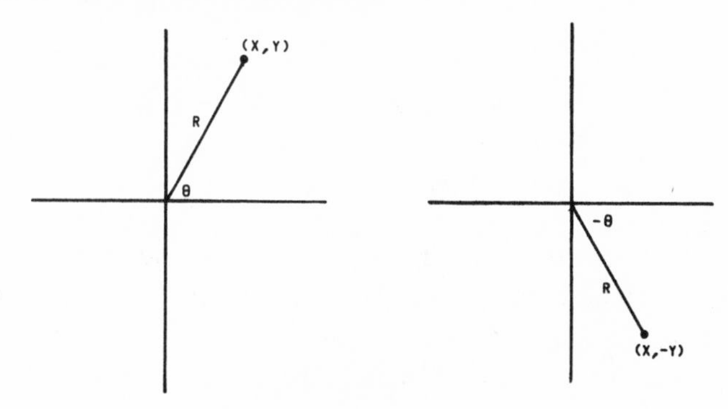

Figure 2-17

As can be seen from the figure, the points in question are defined equally well by their (X,Y) values or by their (R,θ) values. When the points are located by their (R,θ) values, the points are said to be located by their polar values. Since it is possible to express locations in either polar or rectangular coordinates, it must be possible to convert from one system to the other. Figure 2-18 shows how to convert (X,Y) values to (R,θ) values. Point (X,Y) is located by the correct number of units on the X-axis and on the Y-axis. Dotted lines on the figure represent this. From trigonometry and Figure 2-18, θ can be found by:

$$\tan\theta = \frac{Y}{X}$$

$$\theta = \tan^{-1}\left(\frac{Y}{X}\right)$$

Figure 2-18

R can be found by:

$$\sin\theta = \frac{Y}{R} \quad \text{so} \quad R = \frac{Y}{\sin\theta}$$

$$\cos\theta = \frac{X}{R} \quad \text{so} \quad R = \frac{X}{\cos\theta}$$

$$\left\{ \text{R also equals } \sqrt{X^2 + Y^2} \right\}$$

These equations determine R and θ when X and Y are given. It is also possible to convert the other way. In Figure 2-18, if the (R,θ) values are given, (X,Y) values can be found as follows:

$$\sin\theta = \frac{Y}{R} \qquad Y = R\sin\theta$$

$$\cos\theta = \frac{X}{R} \qquad X = R\cos\theta$$

To show how to convert coordinates, consider the equation $Y = MX + B$, where $M = 1$ and $B = 1$. This gives a straight line of slope, 1 and Y intercept 1. This equation can be written in polar form by substituting the conversion equations for X and Y.

$$Y = X + 1 \qquad \text{Substitute} \qquad \begin{array}{l} Y = R\sin\theta \\ X = R\cos\theta \end{array}$$

$$R\sin\theta = R\cos\theta + 1$$

$$\frac{R\sin\theta}{R\cos\theta} = \frac{R\cos\theta + 1}{R\cos\theta}$$

$$\tan\theta = 1 + \frac{1}{R\cos\theta}$$

$$1 = R\cos\theta\,(\tan\theta - 1)$$

$$\boxed{R = \frac{1}{\cos\theta\,(\tan\theta - 1)}}$$

Example 1: Convert $(X,Y) = (2,3)$ to polar

$$\theta = \tan^{-1}\left(\frac{Y}{X}\right) = \tan^{-1}\left(\frac{3}{2}\right) = 56.31^0$$

$$R = \frac{Y}{\sin\theta} = \frac{3}{0.832} = 3.606$$

SECTION 6: HOW TO MANIPULATE EXPONENTS AND RADICALS

Introduction

Equations and other manipulations in electronics often involve the use of exponents and radicals. Equations such as $I = \sqrt{P/R}$ and multiplications such as $(2.3 \times 10^5) \times (3.1 \times 10^4)$ are examples of their use. An understanding of the use of exponents and radicals is necessary when solving electronic problems.

Definition

An exponent is a number to which another number is raised or the number of times a number is multiplied by itself.

For example:

4^3 Equals $4 \cdot 4 \cdot 4 = 64$

In this example, 3 is an exponent of the number four. A radical is a special case of an exponent and is defined by the same laws. The term "radical" usually implies the notation ($\sqrt{\ }$), but it can also be expressed as a fractional exponent.

For example:

$\sqrt{A} = A^{\frac{1}{2}}$ = square root of A

$\sqrt[3]{A} = A^{\frac{1}{3}}$ = cube root of A

$\sqrt[4]{A^3} = A^{\frac{3}{4}}$ = fourth root of A^3

List of Laws

Manipulation of exponents and radicals is done according to the following list of laws:

1. $X^M \cdot X^N = X^{M+N}$
2. $X^M \div X^N = X^{M-N}$
3. $(X^M)^N = X^{MN}$
4. $X^M = \dfrac{1}{X^{-M}}$

5. $(XY)^M = X^M Y^M$

6. $\left(\frac{X}{Y}\right)^M = \frac{X^M}{Y^M}$

7. $X^0 = 1$

8. $X^{M/N} = \sqrt[N]{X^M} = (\sqrt[N]{\ })^M$

Zero, Negative, and Fractional Exponents

Exponents are not limited to real positive numbers. They can have any value. When the exponent is a fraction or a negative number, they can be solved according to the laws listed:

Example 1: $3^{\frac{1}{2}} = \sqrt{3} = 1.732$

Example 2: $5^{-\frac{1}{2}} = \frac{1}{5^{\frac{1}{2}}} = \frac{1}{\sqrt{5}} = \frac{1}{2.236} = 0.4472$

Example 3: $\frac{4 \times 10^{-3}}{2 \times 10^{-2}} = \frac{4}{2} \times \frac{10^{-3}}{10^{-2}} = 2 \times (10^{-3} \times 10^{+2})$

$= 2 \times 10^{-1} = \boxed{0.2}$

Examples of Solved Problems

The following examples represent step-by-step manipulation and simplification of exponents/radicals.

A. $(I^3 R^4)^{\frac{3}{2}} = I^{\frac{9}{2}} R^{\frac{12}{2}} = I^{\frac{9}{2}} R^6$ (or) $\sqrt{I^9} \times R^6$

B. $(27X^2 Y^9 A^8)^{-\frac{1}{3}} = \frac{1}{(27X^2 Y^9 A^8)^{\frac{1}{3}}} = \frac{1}{3X^{\frac{2}{3}} Y^3 X^{\frac{8}{3}}}$

C. $\left(\frac{X^{\frac{1}{3}} Y}{A^{\frac{1}{6}} B}\right)^3 = \frac{X^{\frac{3}{3}} Y^3}{A^{\frac{3}{6}} B^3} = \frac{XY^3}{\sqrt{A}\, B^3}$

D. $10^{-8} \times 10^{+5} = 10^{(-8+5)} = 10^{-3} = \frac{1}{10^3} = \frac{1}{1000} = 0.001$

E. $\frac{10^{-4}}{10^{-5}} = 10^{-4} \times 10^5 = 10^{(-4+5)} = 10^1 = 10$

F. $\frac{1}{\sqrt{2}} = \frac{1}{2^{\frac{1}{2}}} = 2^{-\frac{1}{2}}$

SECTION 7: UNDERSTANDING IMAGINARY NUMBERS

Introduction

Often in mathematics, and especially in electronics math, the situation arises in which we have to take the square root of a negative number $\sqrt{-X}$. Since positive multiplied by positive is positive, and negative multiplied by negative is positive, it is hard to conceive of a number which, when multiplied by itself, turns out negative. Since the square root of a negative number does come up in the solution to electronic problems, it is necessary to learn how to deal with these situations. This section gives examples of how to use these imaginary numbers.

Definition

The ("j" operator) is the symbol used to deal with the square root of negative numbers. It is defined as: $\sqrt{-1}$. Using this definition, the square root of any negative number can be expressed in terms of "j". For example:

$$\sqrt{-4} = \sqrt{4} \times \sqrt{-1} = 2j$$
$$\sqrt{-5} = \sqrt{5} \times \sqrt{-1} = \sqrt{5}\,j = 2.236\,j$$

It is possible to show the relationship of imaginary numbers to real numbers in terms of a graph. First, consider the set of real numbers and how they can be represented on a "number line." Figure 2-19 shows a line with positive numbers on the right-hand side of 0, and negative numbers on the left-hand side.

-9 -8 -7 -6 -5 -4 -3 -2 -1 0 1 2 3 4 5 6 7 8 9

Figure 2-19

This number line represents all real numbers both rational and irrational.

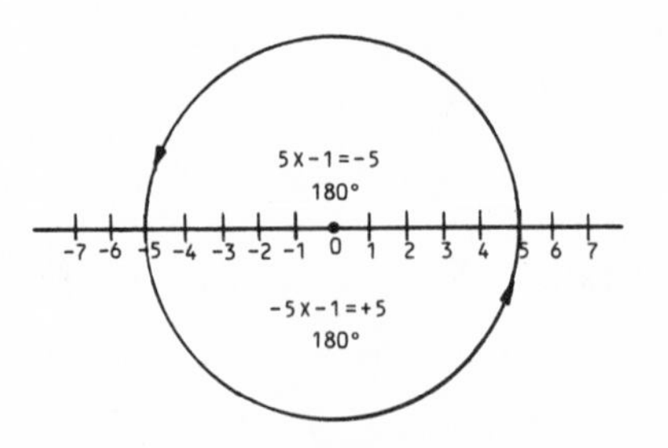

Figure 2-20

Using this number line, consider what happens when a positive number is multiplied by a negative (-1) one. It is shown in Figure 2-20 as a rotation of 180° around zero. If we multiply again by -1, the number is rotated another 180° for a total rotation of 360°, bringing it back to the original number. Multiplication by -1 can be considered as 180° rotation around zero. (See Figure 2-21.)

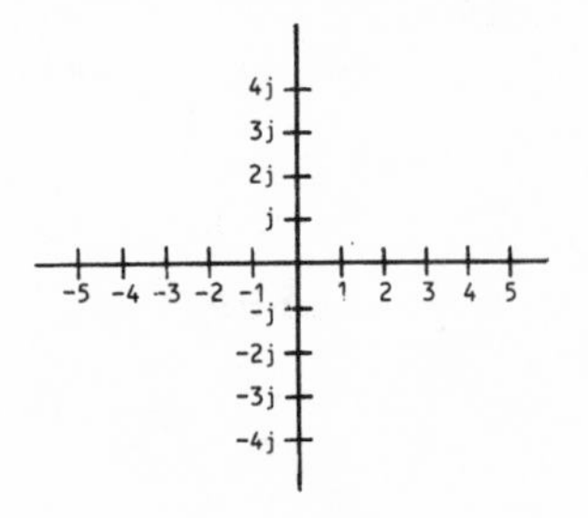

Figure 2-21

Imaginary numbers can also be shown on a number line. The imaginary number line is a line that is perpendicular to the real number line. Since j is defined as $\sqrt{-1}$ and multiplication by -1 can be represented by a 180° rotation, multiplication by j can be indicated graphically, as shown in Figure 2-22 on page 47.

Multiplication of a number by j can be thought of as rotation of the number by 90° or any number (±Aj) that exists on a line that is perpendicular to the real number line.

Complex Numbers

A complex number is a number made from a real number and an imaginary number. In general, complex numbers have the following form:

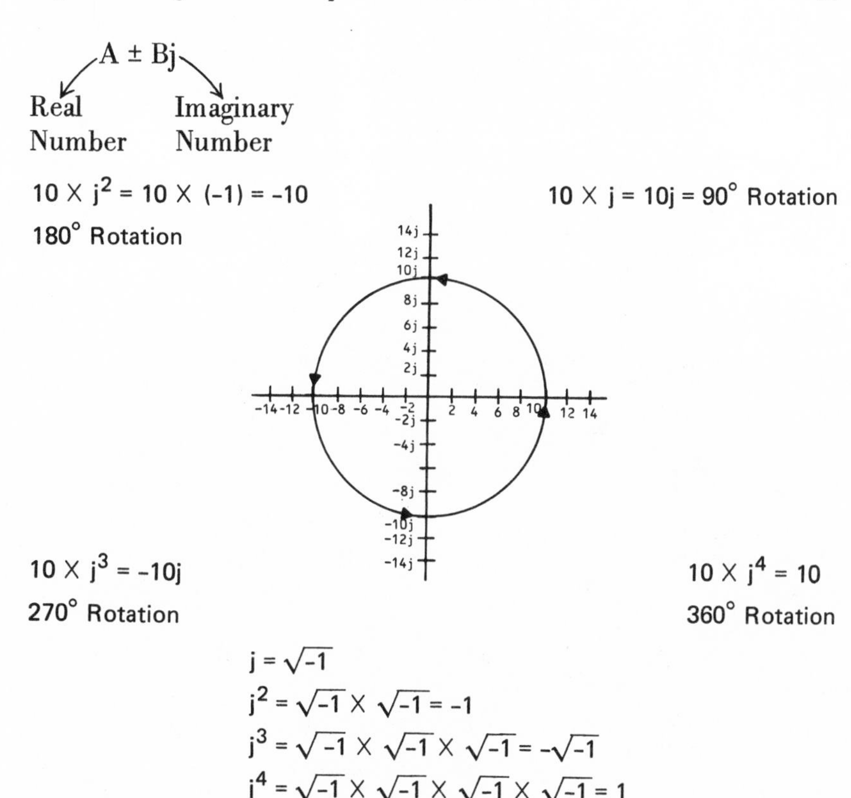

Figure 2-22

Each complex number represents a point on the combined real and imaginary number lines. For example, the complex numbers (5 - 2j) and (4 + 3j) are represented by the points shown in Figure 2-23. Since these complex numbers are represented on two perpendicular lines (rectangular coordinates), they can also be represented in polar form. The conversion to polar form uses the following relationships.

Given the complex number (A + Bj) the angle in polar form is $\text{Angle} = \tan^{-1}\left(\frac{B}{A}\right)$. The magnitude in polar form is $\text{magnitude} = \frac{B}{\sin\theta}$.

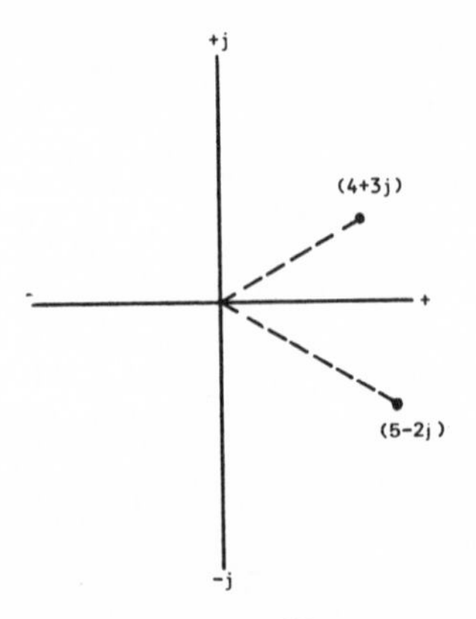

Figure 2-23

Example 1:

Convert 10 - 8j to polar

$$\text{Angle} = \tan^{-1}\left(\frac{-8}{10}\right) = \tan^{-1}(-0.8)$$

$$\tan^{-1}(-0.8) = 38.66^0$$

$$\text{magnitude} = \frac{B}{\sin\theta} = \frac{-8}{\sin(-38.66)} = \frac{-8}{-0.6247} = 12.81$$

$$\text{polar form} = 12.81\ \angle{-38.66}$$

Complex Conjugate

When working with complex numbers, it is often necessary to use a quantity called the complex conjugate. Given a complex number, the complex conjugate is formed by changing the "sign" of the imaginary term. For example:

Complex Number		*Complex Conjugate*
3 + 2j	⟶	3 - 2j
2 - 5j	⟶	2 + 5j
10 + 6j	⟶	10 - 6j

The complex conjugate is useful because, when a complex number is multiplied by its complex conjugate, the imaginary part of the number disappears and all that is left is a real number.

$$\begin{array}{rl} A + Bj & \text{(Complex Number)} \\ \times \quad A - Bj & \text{(Complex Conjugate)} \\ \hline A^2 + \cancel{ABj} - \cancel{ABj} - B^2j^2 & \\ A^2 - B^2j^2 \longrightarrow j^2 = -1 & \\ = A^2 + B^2 & \end{array}$$

Arithmetic with Complex Numbers

The following are examples of addition, subtraction, multiplication, and division using complex numbers.

1. Add: $(8 - 5j)$ and $(-4 + 3j)$

$$\begin{array}{r} 8 - 5j \\ + -4 + 3j \\ \hline 4 - 2j \end{array}$$

2. Multiply $(3 - 2j)$ and $(6 + 2j)$

$$\begin{array}{l} \quad 3 - 2j \\ \times \quad 6 + 2j \\ \hline 18 - 12j + 6j - 4j^2 \\ 18 - 6j - 4j^2 \longrightarrow j^2 = -1 \\ 18 - 6j + 4 \\ \underline{\underline{22 - 6j}} \end{array}$$

3. Subtract: $(4 - 2j)$ from $(5 + 4j)$

$$\begin{array}{rr} 5 + 4j & 5 + 4j \\ -(4 - 2j) & -4 + 2j \\ \hline & = 1 + 6j \end{array}$$

4. Divide: $(1 - 1j)$ by $(2 + 3j)$
   To divide complex numbers, it is necessary to use the complex conjugate.

$$(1 - 1j) \div (2 + 3j) = \frac{1 - 1j}{2 + 3j}$$

$$\frac{1 - 1j}{2 + 3j} = \frac{1 - 1j}{2 + 3j} \times \frac{2 - 3j}{2 - 3j} \longrightarrow$$ Multiply by the complex conjugate.

$$\frac{1 - 1j}{2 + 3j} \times \frac{2 - 3j}{2 - 3j} = \frac{2 - 3j - 2j + 3j^2}{4 + 9}$$

$$\frac{2 - 3j + 2j + 3j^2}{4 + 9} = \frac{-1 - 5j}{13} =$$

$$-0.0769 - 0.3846j$$

SECTION 8: USING SCIENTIFIC NOTATION

Scientific notation is a convenient way to deal with very large or very small numbers. It consists of writing a number with the decimal point after the first or second digit, and then multiplying the number by ten to some power, in order to correct for the decimal point. The value of ten raised to a power is shown in Table 2-1.

0.000001	10^{-6}	ten to the minus 6
0.00001	10^{-5}	ten to the minus 5
0.0001	10^{-4}	ten to the minus 4
0.001	10^{-3}	ten to the minus 3
0.01	10^{-2}	ten to the minus 2
0.1	10^{-1}	ten to the minus 1
1.0	10^{0}	ten to the zero
10.0	10^{1}	ten to the plus 1
100.0	10^{2}	ten to the plus 2
1000.0	10^{3}	ten to the plus 3
10000.0	10^{4}	ten to the plus 4
100000.0	10^{5}	ten to the plus 5
1000000.0	10^{6}	ten to the plus 6

Table 2-1

Consider the small number 0.00000734. It is very cumbersome to write the number in this fashion. It could be written as:

$$(7.34) \times (0.000001)$$

since, when these two numbers are multiplied, the result is the original number. The quantity .0000001 can be written as a power of ten as Table 2-1 shows.

$$0.000001 = 10^{-6}$$

The number can now be written in scientific notation.

$$0.00000734 = (7.34 \times 10^{-6})$$

By the same manner, the number 734000 can be written as follows:

$$734000 = (7.34 \times 10^{+5})$$

When numbers are expressed using scientific notation, the following rules apply.

Rule 1: A negative power of ten says the decimal point should be moved to the left for the same number of places as the power of ten.

Rule 2: A positive power of ten says the decimal point should be moved to the right for the same number of places as the power of ten.

Examples:

$2.3 \times 10^{-3} = 0.0023$

$5.6 \times 10^{+5} = 560{,}000$

$6.3 \times 10^{-9} = 0.0000000063$

Addition and Subtraction

In order to add or subtract numbers that are expressed in scientific notation, it is first necessary to put both numbers to the same power of ten. Consider the following examples:

Example 1: Add 6.3×10^{-4} and 5.2×10^{-5}

Solution: Since the powers of ten must be equal, first convert 6.3×10^{-4} to a 10^{-5} number.

$6.3 \times 10^{-4} = 0.00063$

$0.00063 = 63.0 \times 10^{-5}$

$$
\begin{array}{r}
63.0 \times 10^{-5} \\
+\ 5.2 \times 10^{-5} \\
\hline
= 68.2 \times 10^{-5}
\end{array}
$$

Example 2: Subtract $8.32 \times 10^{+5}$ from $2.83 \times 10^{+6}$.

Solution: Write the numbers so that the powers of ten are equal.

$8.32 \times 10^{+5} = 832000$

$832000. = 0.832 \times 10^{+6}$

$$
\begin{array}{r}
2.830 \times 10^{+6} \\
-\ .832 \times 10^{+6} \\
\hline
= 1.998 \times 10^{+6}
\end{array}
$$

Multiplication and Division

To multiply or divide numbers that are expressed in scientific notation, multiply or divide the numbers without regard to the power of ten, then add the powers of ten if multiplying and subtract the powers of ten if dividing.

Example 1: Multiply $(1.5 \times 10^{+4})(2.3 \times 10^{-2})$

$$
\begin{array}{rl}
1.5 & 10^{+4} \\
\times 2.3 & 10^{-2} \\
\hline
= 3.45 & 10^{+2} = 3.45 \times 10^{+2}
\end{array}
$$

Example 2: Divide $(2.0 \times 10^{+6}) \div (1.5 \times 10^{+4})$

$(2.0) \div (1.5) = 1.333 \qquad 10^{+6} \div 10^{+4} = 10^{(6-4)} = 10^{+2}$

$\underline{1.333 \times 10^{+2}}$

Electronic Problems

It is common in electronic problems to express numbers as powers of ten. When this occurs, it is necessary to perform calculations with these numbers.

Example 1:

In Figure 2-24, the resistance is 10KΩ or $(10 \times 10^{+3})$ ohms, since (K) stands for 10^{+3}. The current is 2.0 MA or (2×10^{-3}) amps, since (milli) stands for 10^{-3}. The voltage is found by Ohm's Law.

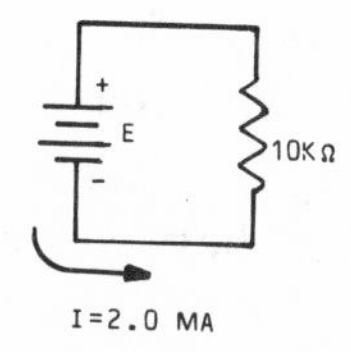

Figure 2-24

$$E = IR = (2.0 \times 10^{-3})(10 \times 10^{+3})$$
$$E = 20 \times 10^{(+3, -3)} = 20 \times 10^{0} = 20 \text{ Volts}$$

SECTION 9: UNDERSTANDING LOGARITHMS

Introduction

The use of logarithms is very common in electronics. It is the heart of decibels, which is a term used throughout the field of electronics. It can also be used to solve complex problems, since it reduces multiplication to addition and division to subtraction.

This section presents all of the laws that govern logarithms and gives several examples of how to work with logarithms in the solution of electronic problems.

Definition

The most simple and most direct definition of a logarithm is: *A logarithm is an exponent.* It is shown in the following example:

$$10^2 = 100$$

The number "2" in the above example is a logarithm. It is usually written as follows:

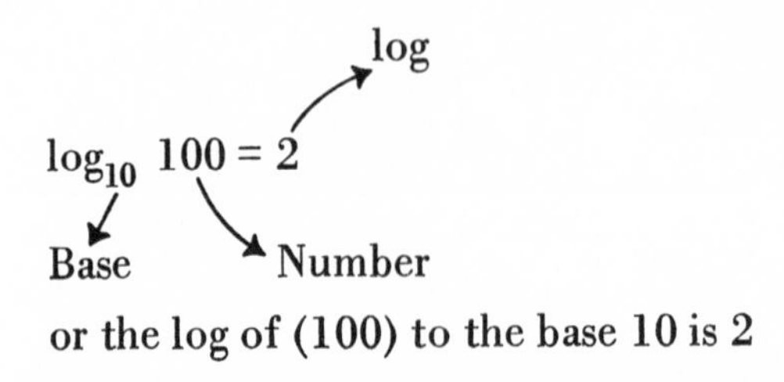

or the log of (100) to the base 10 is 2

The number 10 is used very often as a base of logarithms and logarithms that use this base are called "common" logarithms. It is possible, however, to use any number as a base. The following examples show the relationship between exponential notation and logarithms.

$$2^4 = 16 \longrightarrow \log_2 16 = 4$$
$$5^2 = 25 \longrightarrow \log_5 25 = 2$$
$$10^3 = 1000 \longrightarrow \log_{10} 1000 = 3$$

Common vs. Natural Logs

Logarithms can be defined using any number as a base. In normal usage, however, there are only two numbers used as a base. They are:

base 10 ⟶ common logarithms
base e = 2.71828 ⟶ natural logarithms

Base ten is used because of its relationship to decimal numbers. Base e is used because of its relationship to certain natural events. For example, e is the same e used in the inductive and capacitive time constant curves.

Laws of Logarithms

As with any other branch of algebra, logarithms obey a set of exact laws by which they can be changed and/or solved. A list of these laws is given in Table 2-2.

Antilog and Colog

Antilog and colog are two terms often used when dealing with logarithms. They are defined as follows.

Antilog

If: $\log_{10} 100 = 2$

Then antilog $2 = 100$

Table 2-2

1. $A^X = N - \log_A N = X - N = \text{antilog}_A X$
2. $\log_A A^b = b$
3. $\log_A (M \cdot N) = \log_A M + \log_A N$
4. $\log_A \frac{M}{A} = \log_A M - \log_A N$
5. $\log_A M^N = N \log_A M$
6. $\log_A M^{1/N} = \frac{\log_A M}{N}$
7. $\text{colog}_A N = \log_A \frac{1}{N}$
8. $\log_b A = \frac{1}{\log_A b}$
9. $\log_b N = \log_A N \cdot \log_b A = \frac{\log_A N}{\log_A b}$

The antilog is the reverse operation of logarithms, or if 2 is the common log of 100, then 100 is the antilog of 2.

Colog

A cologarithm is the logarithm of the reciprocal of a number or:

$$\text{colog } x = \log \frac{1}{x}.$$

Also, by using the laws of logarithms,

$$\log \frac{1}{x} = -\log x$$

which says that the colog of a number is a minus log of that same number.

Using Logarithms to Solve Problems

Logarithms can simplify algebraic operations using the following rules:

A. Multiplication is performed by addition of logarithms.

B. Division is performed by subtraction of logarithms.

C. Raising a number to a power is performed by multiplying the logarithm.

Example: Evaluate the following:

$$\frac{20 \times 3^2}{40}$$

First find the logarithms:

$\log_{10} 20 = 1.30103$

$\log_{10} 3 = 0.4771213$

$\log_{10} 40 = 1.60206$

To square (3) take $(\log_{10} 3) \times 2$

$0.4771213 \times 2 = 0.9542426$

Add: $\log_{10} 20 \longrightarrow 0.9542426 + 1.30103 = 2.2552726$

Subtract: $\log_{10} 40 \rightarrow 2.2552726 - 1.60206 = 0.6532126$

The result is the antilog of (0.6532126)

Antilog (0.6532126) = $\boxed{4.5}$

Logarithms in Electronics

Logarithms are used in electronics most commonly in the form of decibels. Decibels are, by definition, the logarithm of the ratio of two quantities such as power, voltage, or current. The basic unit is the *bel* named after Alexander Graham Bell. The definition of the bel is:

$$\text{bel} = \log_{10} \frac{P_1}{P_2}$$

The bel is a very large quantity and is seldom used. The more common term is the *decibel* or one tenth of a bel. The definition of a decibel is:

$$\text{decibel} = 10 \log_{10} \left(\frac{P_1}{P_2}\right)$$

Or, since power is proportional to voltage squared, the equation can be written:

$$\text{decibel} = 10 \log_{10} \frac{E_1^2}{E_2^2}$$

$$\text{decibel} = 20 \log_{10} \frac{E_1}{E_2}$$

The use of the decibel is best understood by specific examples of their use.

Example 1:

The amplifier in Figure 2-25 has an output power of 10 watts, with an input of 0.1 watts. What is the amplifier gain in decibels?

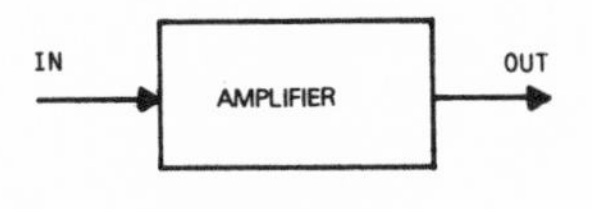

Figure 2-25

$$\text{db} = 10 \log_{10} \frac{P_{out}}{P_{in}}$$

$$= 10 \log_{10} \frac{10}{0.1} = 10 \log_{10} (100)$$

$$\text{db} = 10 \times \log_{10} 100, \quad \log_{10} 100 = 2$$

$$\text{db} = 10 \times 2 = 20$$

Example 2:

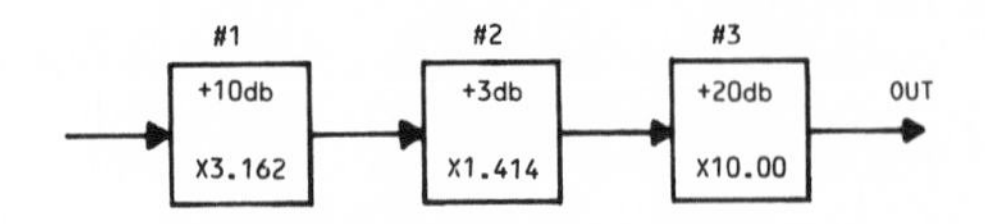

Figure 2-26

Figure 2-26 shows a three-stage cascaded amplifier. The gain of each stage is shown in decibels and in voltage gain. When amplifiers are connected in this manner, their gains are multiplied so that the total gain is:

$$(\text{Gain } 1) \times (\text{Gain } 2) \times (\text{Gain } 3) = \text{Total Gain}$$

Since logarithms reduce multiplication to addition, the total gain can also be expressed in decibels as:

$$(\text{Gain } 1) + (\text{Gain } 2) + (\text{Gain } 3) = \text{Total Gain}$$

The two methods produce the same results as shown below.

$$\text{Total Gain} = 1 \times 2 \times 3$$

$$= 3.162 \times 1.414 \times 10$$

$$= 44.7$$

$$\text{Total Gain} = db_1 + db_2 + db_3$$

$$= 10 + 3 + 20 = 33 \text{ db}$$

$$33 = 20 \log_{10} \frac{E_{out}}{E_{in}}$$

$$\log_{10} \frac{E_{out}}{E_{in}} = \frac{33}{20} = 1.65$$

$$\frac{E_{out}}{E_{in}} = 10^{1.65} = 44.7$$

Example 3:

The effective radiated power (ERP) of a broadcast station is equal to the power applied to the antenna, times the gain of the antenna. If 1000 watts are applied to the antenna and the antenna has a gain of 5.1 db, what is the ERP?

Gain of antenna = 5.1 db

$$5.1 = 10 \log_{10} \frac{P_{out}}{P_{in}} \qquad \log_{10} \frac{P_{out}}{P_{in}} = 0.51$$

$$10^{\left(\log_{10} \frac{P_{out}}{P_{in}}\right)} = 10^{0.51}$$

$$\frac{P_{out}}{P_{in}} = 3.236$$

$$\text{ERP} = 1000 \times 3.236 = \underline{3236 \text{ watts}}$$

Reference Levels

The decibel, by itself, is simply a ratio of two levels. There is no reference level with which to compare the numbers. Several different reference levels have been defined and are used in electronics. Some of the more common ones are given:

V.U. (Volume Units) ⟶ Zero V.U. = 0.001 watts Across a 600Ω Load

dbm ⟶ Zero dbm can mean either 0.001 watts or 0.006 watts

Calculations using reference levels are the same as the previous examples using decibels.

Example 4:

10 V.U. (volume units) represents how much power?

$$10 \log_{10} \frac{P_{out}}{0.001} = 10$$

$$\log_{10} \frac{P_{out}}{0.001} = 1$$

$$10^{\left(\log_{10} \frac{P_{out}}{0.001}\right)} = 10^1$$

$$\frac{P_{out}}{0.001} = 10$$

$$P_{out} = 0.01 \text{ watts}$$

3 HOW TO SOLVE PASSIVE ELECTRONIC CIRCUITS

Solving circuits is basic to understanding electronics. This chapter defines the fundamental laws involved and gives several examples of how to use these laws.

QUICK REFERENCE LIST OF LAWS AND FORMULAS

Electronic circuits in series and parallel containing voltage sources, current sources, resistors, inductors, and capacitors can be solved using the following laws:

1. *Ohm's Law*—The current through a circuit element is directly proportional to the applied voltage and inversely proportional to the resistance. (See Figure 3-1.)

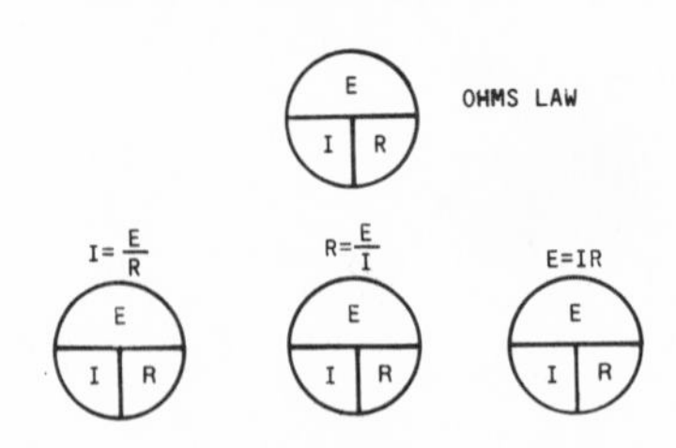

Figure 3-1

2. *Kirchoff's Voltage Law*—The algebraic sum of the voltage drops around any loop equals zero. (See Figure 3-2.)

$$V_1 + V_2 + V_3 + V_4 = 0$$

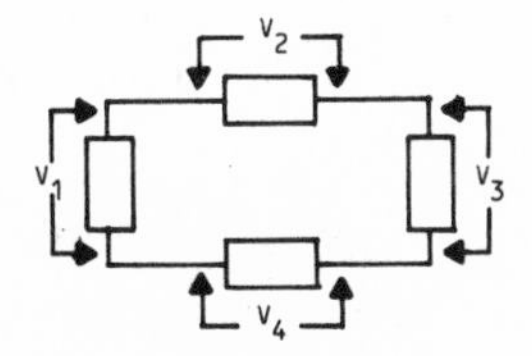

Figure 3-2

3. *Kirchoff's Current Law*—The algebraic sum of currents at a node* equals 0. (See Figure 3-3.)

$$I_1 + I_2 + I_3 = 0$$

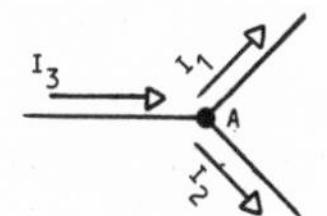

Figure 3-3

4. *Thevenin Theorem*—A network of power sources and components can be reduced to a single voltage source and a series resistor. (See Figure 3-4.)

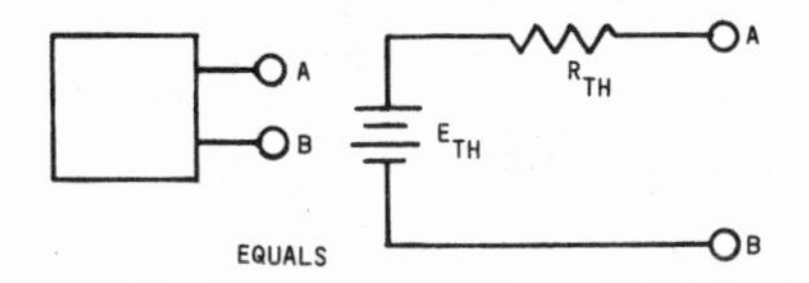

Figure 3-4

5. *Norton's Theorem*—A network of power sources and components can be reduced to a single current source and a parallel resistor. (See Figure 3-5.)

---

*A node is defined as the junction of two or more current paths.

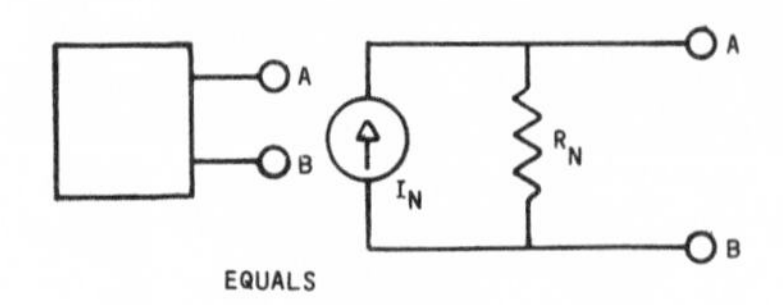

Figure 3-5

6. *Superposition*—In an electronic circuit with two or more power sources, the current, or voltage for any component is the algebraic sum of each source acting separately.

To determine the effects of one source, the other sources are disabled. Voltage sources are considered shorts and current sources open.

EXAMPLES OF SOLUTIONS USING OHM'S LAW

Ohm's law is the most fundamental law in electronics. It relates voltage (electronic pressure) to current (amount of electron flow) and resistance (opposition to electron flow).

Stated as $I = \frac{E}{R}$, it says that the amount of electron flow is directly proportional to the pressure (voltage) applied and inversely proportional to the resistance (R). For example, if the pressure is doubled, the current will double. If the resistance is doubled, the current will be reduced to one-half.

A. Single Resistor. (See Figure 3-6.)

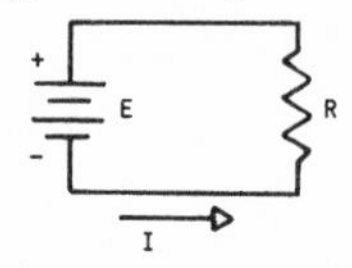

Figure 3-6

1. Given: E = 10 volts, R = 5KΩ, find I

$$I = \frac{E}{R} = \frac{10}{5000} = 0.002 \text{ amps}$$

2. Given: E = 100 volts, I = 0.05 amps, find R

$$R = \frac{E}{I} = \frac{100}{.05} = 2000\Omega$$

3. Given: $R = 10K\Omega$, $I = 0.47$ amps, find E

$$E = IR = 10{,}000 \times 0.47 = 4700 \text{ volts}$$

B. Series Resistors. (See Figure 3-7.)

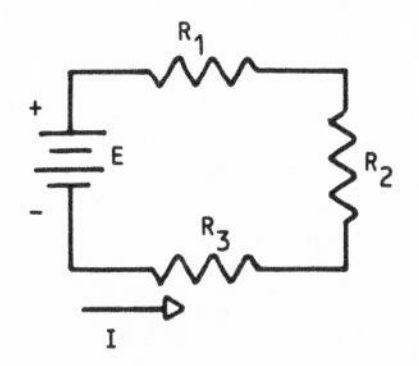

Figure 3-7

In a series circuit, the total resistance: $R_T = R_1 + R_2 + R_3 + \ldots.$

1. Given: $E = 100V$, $R_1 = 1K\Omega$, $R_3 = 2K\Omega$, $R_3 = 3K\Omega$, find I

$$I = \frac{E}{R_T} = \frac{E}{R_1 + R_2 + R_3} = \frac{100}{6000} = 0.0166 \text{ amps}$$

2. Given: $I = 0.1$ amps, $E = 200$ volts, $R_1 = 500\Omega$, $R_2 = 700\Omega$, find R_3

$$R_T = \frac{E}{I} = \frac{200}{0.1} = 2000\Omega$$

$$R_3 = R_T - (R_1 + R_2) = 2000 - (700 + 500)$$

$$R_3 = 800\Omega$$

3. Given: $I = 1A$, $R_1 = 5\Omega$, $R_2 = 2\Omega$, $R_3 = 5.5\Omega$, find E

$$E = IR_T = 1(5 + 2 + 5.5) = 12.5V$$

C. Parallel Resistors. (See Figure 3-8.)

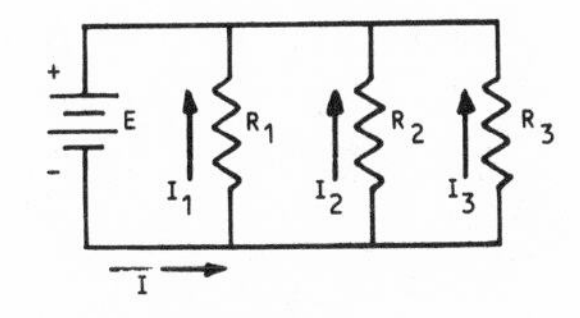

Figure 3-8

In a parallel circuit, the total resistance

$$R_T = \frac{1}{\frac{1}{R_1} + \frac{1}{R_2} + \frac{1}{R_3} + \ldots}$$

$$R_T = \frac{R_1 R_2}{R_1 + R_2} \quad \text{in the case of two resistors}$$

1. Given: $E = 50$ volts, $R_1 = 100\Omega$, $R_2 = 50\Omega$, $R_3 = 150\Omega$, find I_T, and I_2

$$I_2 = \frac{E}{R_2} = \frac{50}{50} = 1 \text{ amp}$$

$$I_T = \frac{E}{R_T} \qquad R_T = \frac{1}{\frac{1}{R_1} + \frac{1}{R_2} + \frac{1}{R_3}}$$

$$R_T = \frac{1}{.01 + .02 + .00666} = 27.27\Omega$$

$$I_T = \frac{E}{R_T} = \frac{50}{27.27} = 1.833 \text{ amps}$$

D. Series Parallel Circuits: (See Figure 3-9.)

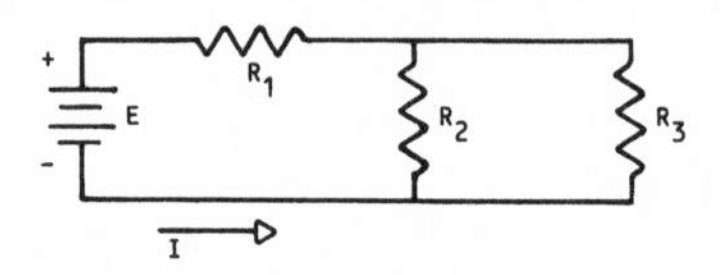

Figure 3-9

1. Given: $E = 100$ volts, $R_1 = 2K\Omega$, $R_2 = 5K\Omega$, $R_3 = 3K\Omega$, find I

$$I = \frac{E}{R_T} \qquad R_T = R_1 + \text{parallel } R_2, R_3$$

$$R_2, R_3 = \frac{R_2 R_3}{R_2 + R_3} = \frac{5K \times 3K}{8K} = 1.875K\Omega$$

$$R_T = 2K + 1.875K = 3.875K$$

$$I = \frac{E}{R_T} = \frac{100}{3875} = 0.0258 \text{ amps}$$

2. Using Example 1, find the voltage across R_1:

$$E_{R_1} = R_1 \times I_T = 2000 \times 0.0258$$
$$= 51.6 \text{ volts}$$

EXAMPLES OF SOLUTIONS USING KIRCHOFF'S LAWS

Kirchoff's laws deal with current nodes and voltage loops. They allow sufficient equations to be written to solve any circuit.

The current law says that the amount of current entering a node (junction) must equal the amount of current leaving the node. (See Figure 3-10.) If this were not true, a large charge would be building at the junction, or the charge would be depleting at the junction. This cannot happen.

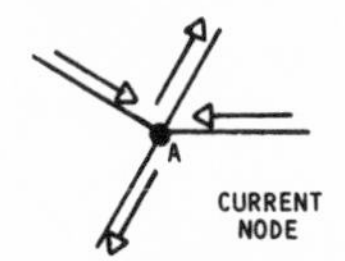

Figure 3-10

The voltage law says that the drops around any loop must add to zero. (See Figure 3-11.) In terms of electric pressure, if you start at one point and go around any loop, you must be back at the same pressure.

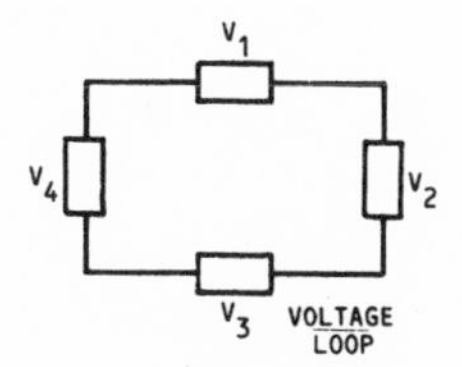

Figure 3-11

Example 1:

1. Given: $E_2 = 10V$, $E_1 = 5V$, $R_1 = 12\Omega$, $R_2 = 10\Omega$, $R_3 = 5\Omega$, find the voltage across R_3. (See Figure 3-12.)

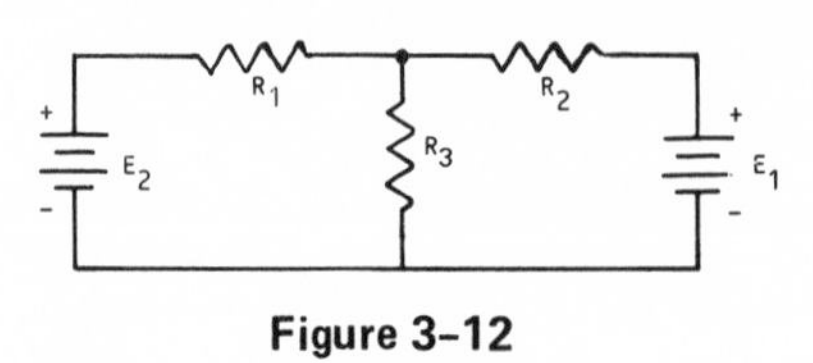

Figure 3-12

First, establish the current nodes and voltage loops. (See Figure 3-13.) The currents I_1, I_2, and I_3 are given a direction arbitrarily. If they are wrong, their values will be negative when Kirchoff's Laws are solved.

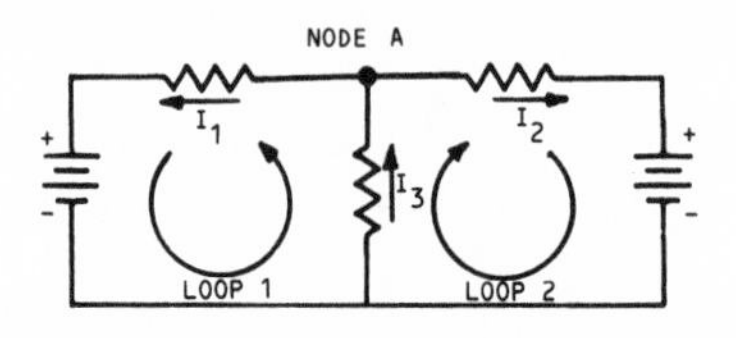

Figure 3-13

Equation 1:

Current Equation at A:

$I_3 = I_2 + I_1$ Equation 1

Equation 2:

Voltage Equation (Loop 1):

$10 - VR_1 - VR_3 = 0$

$VR_1 = 12I_1$

$VR_3 = 5I_3$

$10 - 12I_1 - 5I_3 = 0$ Equation 2

Voltage Equation (Loop 2):

$5 - VR_2 - VR_3 = 0$

$VR_2 = 10I_2$

$VR_3 = 5I_3$

$5 - 10I_2 - 5I_3 = 0$ Equation 3

Equations 1, 2, and 3 represent three independent equations in three unknowns and therefore can be solved.

Equation 3:

$$5 - 10I_2 - 5I_3 = 0$$

Substituting ($I_2 = I_3 - I_1$) from Equation 1

$$5 - 10(I_3 - I_1) - 5I_3 = 0$$

$$5 - 15I_3 - 10I_2 = 0 \qquad \text{Equation 4}$$

Multiplying Equation 2 by (-3):

$$-30 + 36I_1 + 15I_3 = 0$$

Subtracting from Equation 4:

$$\begin{array}{rrrl} 5 & - 15I_3 & - 10I_1 & = 0 \\ -30 & + 15I_3 & + 36I_1 & = 0 \\ \hline -25 & & + 26I_1 & = 0 \end{array}$$

$$I_1 = \frac{25}{26} = 0.9615 \text{ amps}$$

Therefore:

$$V_{R3} = R_3 I_1 = 5 \times 0.9615 = 4.807 \text{ volts}$$

Example 2: Kirchoff's Laws: (See Figures 3-14 and 3-15.)

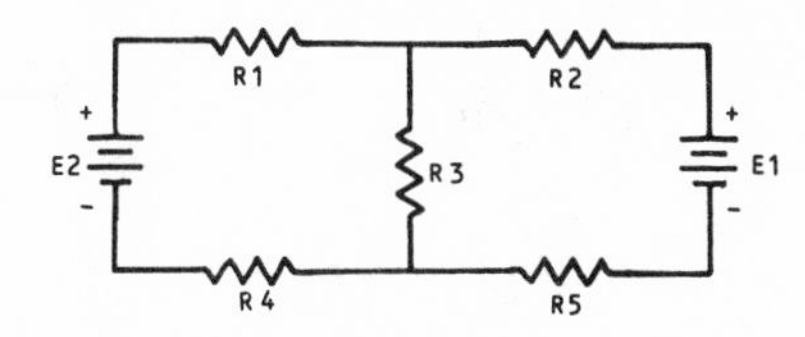

Figure 3-14

Given: $E_2 = 25V$, $E_1 = 10V$, $R_1 = 10\Omega$, $R_2 = 15\Omega$, $R_3 = 10\Omega$, $R_4 = 8\Omega$, $R_5 = 12\Omega$

Find: The voltage across R_2

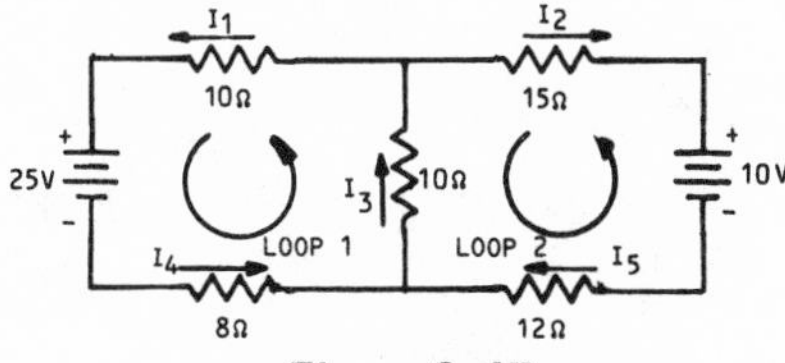

Figure 3-15

$$I_1 = I_4$$
$$I_2 = I_5$$

Voltage around Loop 1:

$$25 - 10I_1 - 8I_1 - 10I_3 = 0$$
$$25 - 18I_1 - 10I_3 = 0 \qquad \text{Equation 1}$$

Voltage around Loop 2:

$$10 - 12I_2 - 10I_3 - 15I_2 = 0$$
$$10 - 27I_2 - 10I_3 = 0 \qquad \text{Equation 2}$$

Current at Node A:

$$I_3 = I_1 + I_2$$
$$\text{or} \quad I_1 = I_3 - I_2 \qquad \text{Equation 3}$$

Substituting Equation 3 in Equation 1:

$$25 - 18(I_3 - I_2) - 10I_3 = 0$$
$$25 - 28I_3 - 18I_2 = 0 \qquad \text{Equation 4}$$

Multiply Equation 2 by 2.8 and Subtract from Equation 4:

$$\begin{array}{rrrl} 25 & - 28I_3 & - 18I_2 & = 0 \\ -28 & + 28I_3 & + 75.6I_2 & = 0 \\ \hline -\ 3 & & + 57.6I_2 & = 0 \end{array}$$

$$I_2 = \frac{3}{57.6} = .0521 \text{ amps}$$

$$V_{R2} = I_2 R_2 = .0521 \times 15 = \boxed{0.781 \text{ volts}}$$

Substitute I_2 into Equation 2:

$$10 - 27(.0521) - 10I_3 = 0$$
$$I_3 = 0.85933 \text{ amps}$$

From Equation 3:

$$I_1 = I_3 - I_2$$
$$= .85933 - .0521$$
$$I_1 = .80723 \text{ amps}$$

EXAMPLES OF CIRCUIT SOLUTIONS USING THEVENIN AND NORTON

Thevenin and Norton Theorems are useful tools in circuit analysis since they allow the simplification of complex circuits. Thevenin reduces a network of power sources and resistors to a single voltage source with a series resistor. Norton reduces the same network to a single current source with a parallel resistor.

Thevenin Theorem (*Example 1*):

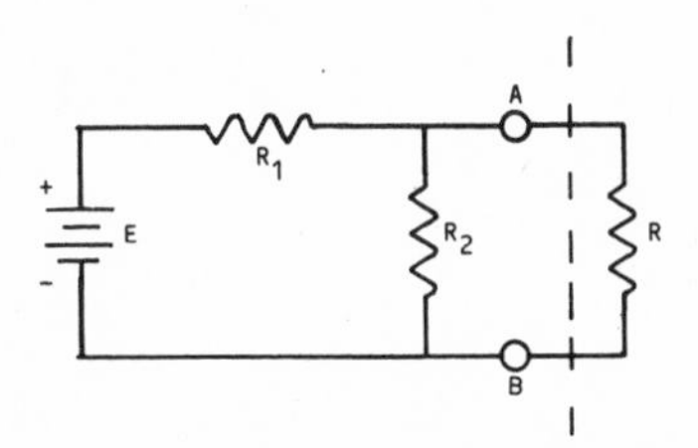

Figure 3-16

(See Figure 3-16.)

E, R_1, and R_2 make up a network of power sources and resistors. R can be considered a load resistor connected to this network at terminals A and B.

Given: $E = 20V$, $R_1 = 10\Omega$, $R_2 = 15\Omega$, $R = 5\Omega$

Find: The current through R.

Method: Reduce E, R_1, and R_2 to a voltage source with a single series resistor and redraw the circuit. (See Figure 3-17.)

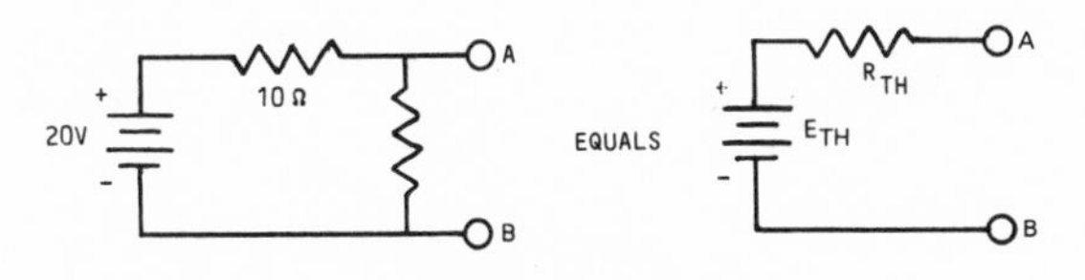

Figure 3-17

To find the Thevenin voltage (E_{TH}), open circuit the load and determine the voltage. (See Figure 3-18.)

$$E_{AB} = 20 \times \left(\frac{15}{10 + 15}\right) = 12 \text{ volts}$$

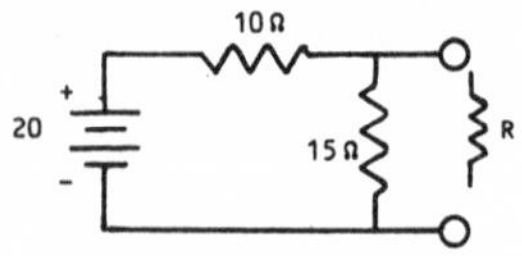

Figure 3-18

To find the Thevenin resistance (R_{TH}), short circuit the terminals A and B, and determine the current through the short. (See Figures 3-19 and 3-20.)

$$R_{TH} = \frac{E_{TH}}{I_{Short}}$$

$$I = \frac{E}{R} = \frac{20}{10} = 2 \text{ amps}$$

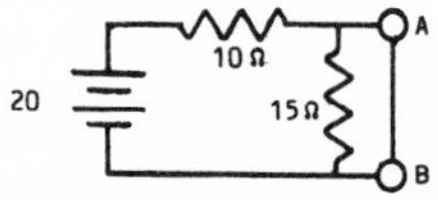

Figure 3-19

$$I = 2 \text{ amps}$$

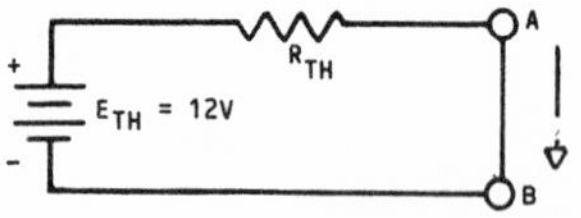

Figure 3-20

$$R_{TH} = \frac{E_{TH}}{I} = \frac{12}{2} = 6 \text{ ohms}$$

The circuit redrawn with the Thevenin equivalent looks like Figure 3-21.

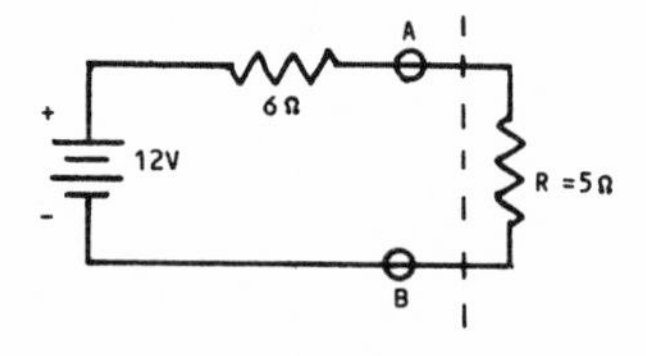

Figure 3-21

The current through R: $I = \frac{12}{11} = 1.09$ amps

Example 2:

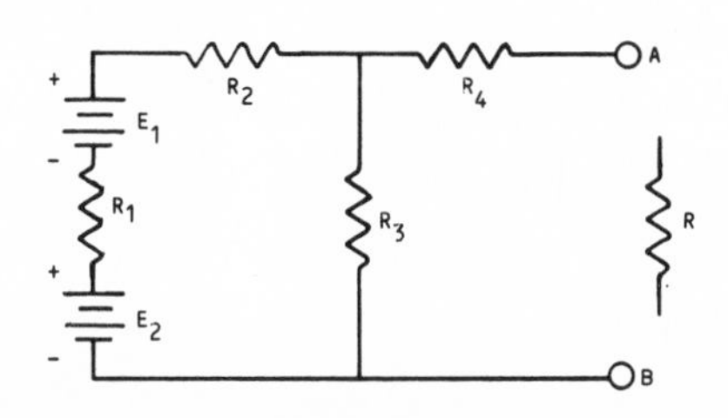

Figure 3-22

E_1, E_2, R_1, R_2, R_3, and R_4 make up a network of power sources and resistors that can be reduced to a single voltage source and a single series resistor. (See Figure 3-22.)

Given: $E_1 = 25V$, $E_2 = 15V$, $R_1 = 5\Omega$, $R_2 = 8\Omega$, $R_3 = 12\Omega$ $R_4 = 10\Omega$, $R = 11\Omega$.

Find: The current through R.

Method: Reduce the complex circuit to its Thevenin equivalent. (See Figure 3-23.)

E_{TH} = Open Circuit Voltage = Voltage drop across R_3

$$I = \frac{E_1 + E_2}{R_1 + R_2 + R_3} = \frac{40}{25} = 1.6 \text{ amps}$$

$$V_{R3} = 1.6 \times 12 = 19.2 \text{ volts} = E_{TH}$$

$$R_{TH} = \frac{E_{TH}}{I_{\text{Short Circuit}}}$$

$$R_4 \| R_3 = \frac{12 \times 10}{12 + 10} = 5.455\Omega$$

$$I_2 = \frac{E_1 + E_2}{5 + 8 + 5.455} = \frac{40}{18.455}$$

$I_2 = 2.167$ amps

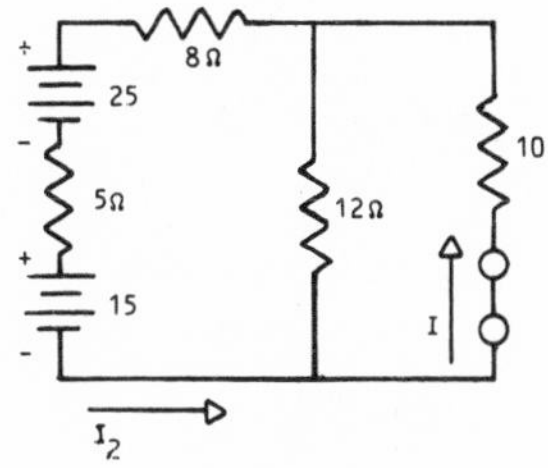

Figure 3-23

Voltage across 10Ω $R_4 = E_{Total} - E_{R_1} - E_{R_2}$

Voltage $= 40 - (2.167)(5) - (2.167)(8)$

Voltage = 11.827 volts

$$I_{Short} = \frac{11.827}{10} = 1.1827 \text{ amps}$$

$$R_{TH} = \frac{E_{TH}}{I_{Short}} = \frac{19.2}{1.1827} = 16.234\Omega$$

With the circuit redrawn: (See Figure 3-24.) The current through R:

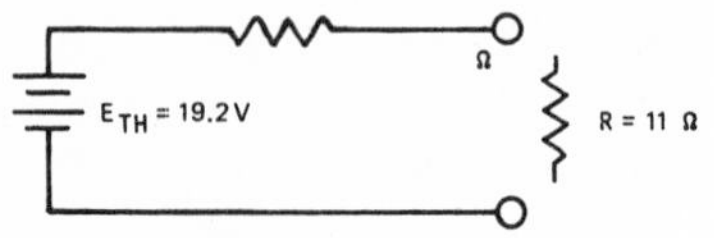

Figure 3-24

$$I = \frac{19.2}{27.234} = \underline{0.705 \text{ amps}}$$

NORTON'S THEOREM

Norton's theorem, like Thevenin, states that a complex circuit of power sources and resistors can be reduced. The Norton reduction gives a current source in parallel with a resistor instead of the Thevenin voltage source and a series resistor. The two are equivalent, so that when the Thevenin circuit is known, the Norton circuit can be determined and, given the Norton circuit, the Thevenin circuit can be determined (See Figure 3-25.)

Example 1:

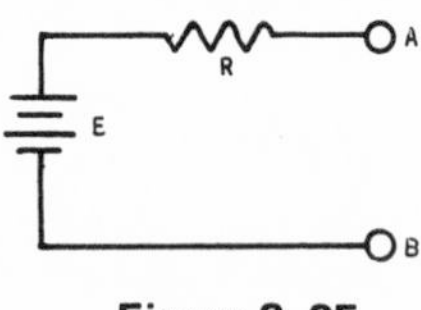

Figure 3-25

Given: E = 10 volts, R = 18Ω

Find: The Norton Circuit

The Norton Circuit looks like Figure 3-26:

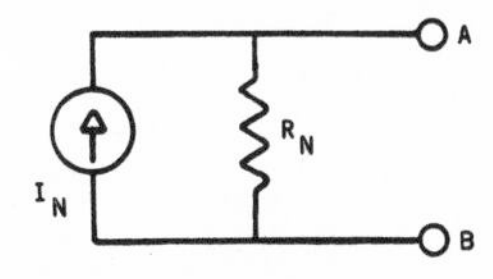

Figure 3-26

I_N = The Short Circuit (A to B) Current

$$I_N = \frac{E}{R} = \frac{10}{18_R} = 0.5555 \text{ amps}$$

R_N is determined by the open circuit (A to B) voltage, in Figure 3-27, 10 volts:

$$R = \frac{E}{I} = \frac{10}{.5555} = 18.00\Omega$$

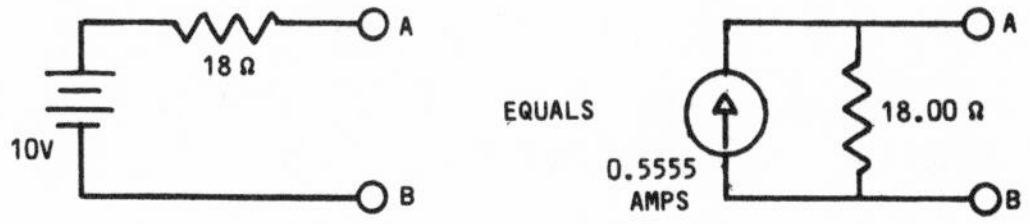

Figure 3-27

The Thevenin (R) and the Norton (R) are always the same value.

Example 2:

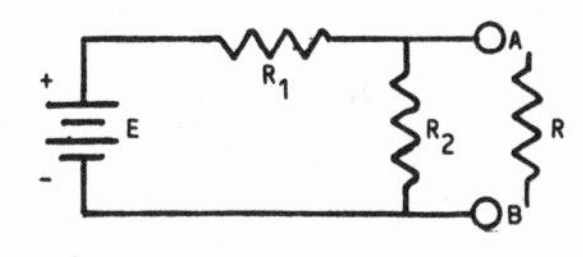

Figure 3-28

Given: $E = 25V$, $R_1 = 10\Omega$, $R_2 = 15\Omega$, $R = 10\Omega$

Find: The voltage across R

Method: Reduce the circuit to its Norton equivalent. (See Figure 3-29.)

I_N = Short Circuit (A to B) Current

$$I_N = \frac{25}{10} = 2.5 \text{ amps}$$

$$R_N = \frac{V_{open}}{I_N} = \frac{15}{2.5} = 6\Omega$$

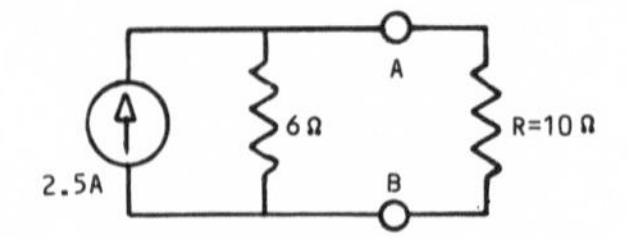

Figure 3-29

Voltage across R = 2.5A × (Total Resistance)

$$V = (2.5) \times \frac{10 \times 6}{10 + 6} = 9.375 \text{ volts}$$

EXAMPLES OF CIRCUIT SOLUTIONS USING SUPERPOSITION

The current or voltage for any component in a circuit can be determined by the algebraic sum of the voltages or currents produced by each source. (See Figure 3-30.)

Example 1:

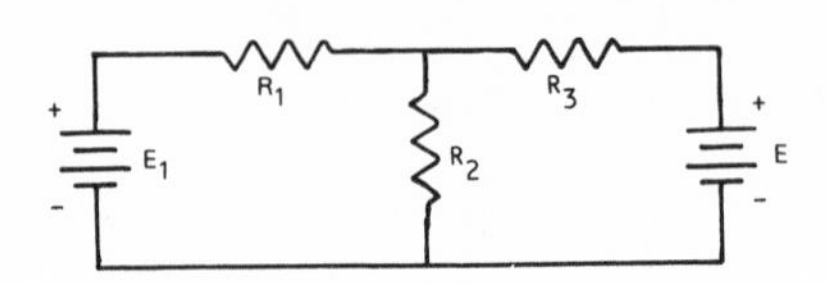

Figure 3-30

Given: $E_1 = 20V$, $E_2 = 30V$, $R_1 = 30\Omega$, $R_2 = 20\Omega$, $R_3 = 30\Omega$

Find: The current through R_3.

Method: The current through R_3 is the algebraic sum of the current produced independently by E_1 and E_2.

The current produced by E_1 is found by drawing the circuit with E_2 shorted, as in Figure 3-31.

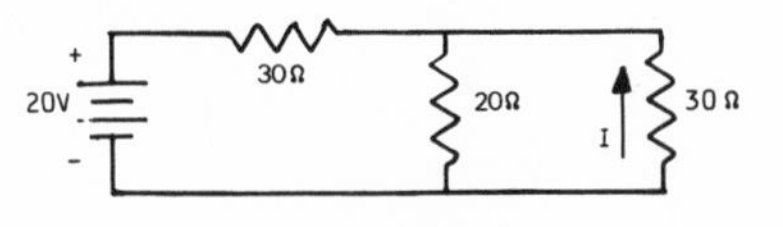

Figure 3-31

$$R_2 \| R_3 = \frac{20 \times 30}{20 + 30} = 12\Omega$$

$$V_{R3} = 20 \times \frac{12}{42} = 5.714$$

$$I_{R3} = \frac{VR_3}{R} = \frac{5.714}{30} = 0.190 \text{ amps}$$

The current produced by E_2 is found by drawing the circuit with E_1 shorted, as in Figure 3-32.

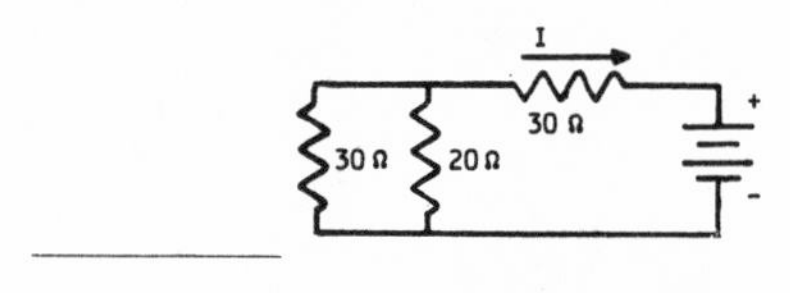

Figure 3-32

$$I_{R3} = \frac{30}{R_T} \qquad R_T = 30 + R_1 \| R_2$$

$$R_1 \| R_2 = \frac{30 \times 20}{30 + 20} = 12\Omega$$

$$R_T = 42\Omega$$

$$I = \frac{30}{42} = 0.714 \text{ amps}$$

The two currents flow through R_3 in opposite directions. The total current, therefore, is the difference between the two. The current direction is the direction of the largest current. (See Figure 3-33.)

$$I_{R3} = 0.714 - 0.190 = 0.524 \text{ amps}$$

Example 2:

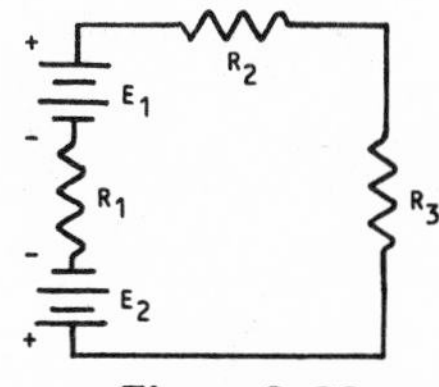

Figure 3-33

Given: $E_1 = 50V$, $E_2 = 20V$, $R_1 = 10\Omega$, $R_2 = 25\Omega$, $R_3 = 48\Omega$
Find: I
I due to E_1 is found by shorting E_2. (See Figure 3-34.)

$$I = \frac{50}{10 + 48 + 25} = 0.602A$$

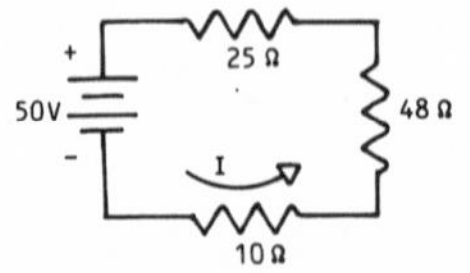

Figure 3-34

I due to E_2 is found by shorting E_1. (See Figure 3-35.)

$$I = \frac{20}{10 + 25 + 48} = 0.241 \text{ amps}$$

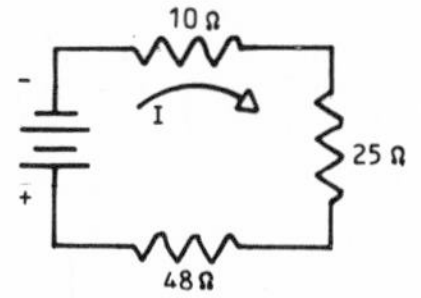

Figure 3-35

$$I_{Total} = I_{E_1} - I_{E_2} = 0.602 - 0.241 = 0.361 \text{ amps}$$

Example 3:

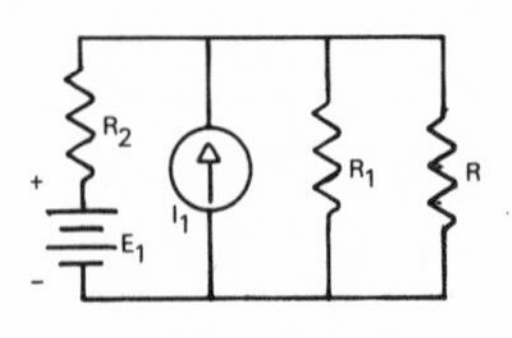

Figure 3-36

Given: $E_1 = 10V$, $I = 1A$, $R_2 = 5\Omega$, $R_1 = 8\Omega$, $R = 10\Omega$
Find: The current through R. (See Figure 3-36.)

The current due to E_1 is found by opening I_1. (See Figure 3-37.)

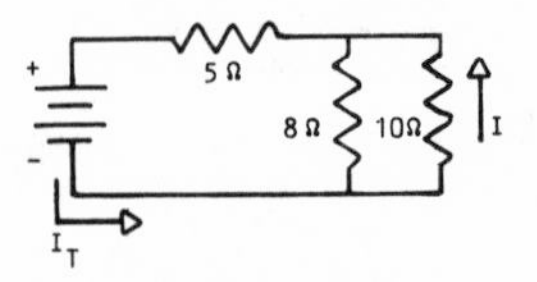

Figure 3-37

$$R \| R_1 = \frac{8 \times 10}{8 + 10} = 4.444\Omega$$

$$I_T = \frac{10}{9.444} = 1.059 \text{ amps}$$

$$V_{5\Omega} = IR = 5 \times 1.059 = 5.295 \text{ volts}$$

$$V_R = 10 - 5.295 = 4.705 \text{ volts}$$

$$I_R = \frac{E}{R} = \frac{4.705}{10} = 0.4705 \text{ amps}$$

The current due to I_1 is found by shorting E_1. (See Figure 3-38.)

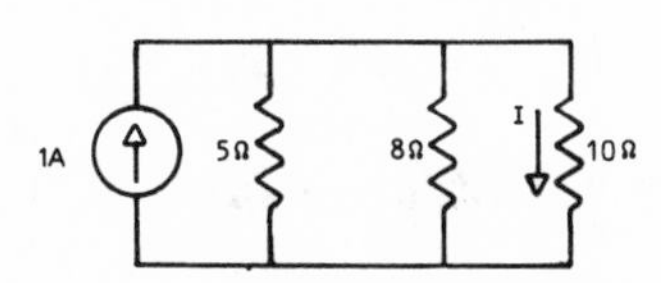

Figure 3-38

$$R_T = R_1 \| R_2 \| R_3 \qquad R_2 \| R_1 = \frac{5 \times 8}{5 + 8} = 3.077\Omega$$

$$3.077 \| 10 = \frac{30.77}{13.077} = 2.353\Omega$$

$$E_T = IR_T = 1 \times 2.353 = 2.353 \text{ volts}$$

$$I_R = \frac{E_R}{R} = \frac{2.353}{10} = 0.2353 \text{ amps}$$

$$I_R = 0.4705 - 0.2353 = 0.2352 \text{ amps}$$

HOW TO SOLVE COMPLEX A.C. CIRCUITS

The same laws used to solve D.C. circuits are used to solve A.C. circuits. The calculations become more involved since the reactive elements (inductors and capacitors) cause different phase relationships between the voltage and currents.

Capacitors

Capacitors obey Ohm's Law when an A.C. voltage is applied. The opposition to current is called capacitive reactance and is given the symbol X_c. X_c is obtained from the following formula:

$$X_c = \frac{1}{2\pi fc} = \frac{0.159}{fc}$$

Example:

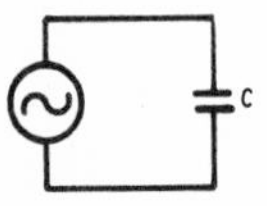

Figure 3-39

Given: E = 10V @ 1MHZ, c = 0.1μFD or 1.0×10^{-7} Farads
Find: I

$$X_c = \frac{0.159}{10^6 \times 0.1 \times 10^{-6}} = 1.59\Omega$$

$$I = \frac{V}{X_c} = \frac{10}{1.59} = 6.289 \text{ amps}$$

The voltage and current across any capacitor are 90° out of phase. The current leads the voltage. (See Figure 3-39.)

Inductors

Inductors obey Ohm's Law when an A.C. voltage is applied. The Ohmic value is called inductive reactance and is given the symbol X_L. X_L is obtained from the following formula:

$$X_L = 2\pi fL = 6.28fL$$

Example:

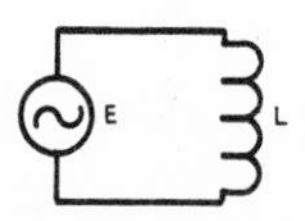

Figure 3-40

Given: E = 10V, @10MHZ, L = 1μH or 1.0×10^{-6} henries

Find: I

$$X_L = 6.28fL = 6.28 \times 10^7 \times 10^{-6} = 62.8 \text{ ohms}$$

$$I = \frac{E}{X_L} = \frac{10}{62.8} = 0.159 \text{ amps}$$

The voltage and current across an inductor are 90° out of phase. The voltage leads the current. (See Figure 3-40.)

Series Circuits

When resistors, capacitors, or inductors are wired in series, the total opposition to current flow is called impedance (Z). The reactance and resistance add as vectors. See the example in Figure 3-41.

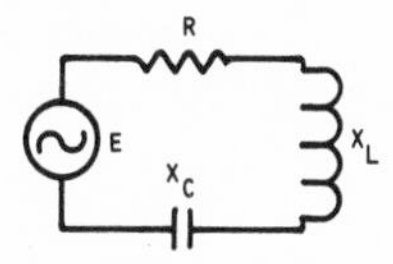

Figure 3-41

Given: E = 20V, R = 20Ω, X_L = 50Ω, X_c = 10Ω

Find: I, E_R, E_L, E_c

The current in this circuit is found by Ohm's Law, $I = \frac{E}{Z}$

Where $Z = \sqrt{(X_L - X_c)^2 + R^2}$

$$Z = \sqrt{40^2 + 20^2} = \sqrt{2000} = 44.72\Omega$$

$$I = \frac{20}{44.72} = 0.447 \text{ amps}$$

Voltage across R = IR = 0.447×20 = 8.94 volts

Voltage across L = IX_L = 0.447×50 = 22.35 volts

Voltage across C = IX_c = 0.447×10 = 4.47 volts

It should be noted that the sum of the voltage drops $V_r + V_L + V_c$ is greater than the source voltage. These voltage drops are vectors and, when added as such, will equal the source voltage.

Parallel Circuits

When resistors, capacitors, and inductors are wired in parallel, the impedance (Z) is determined by the reciprocals of reactance and resistance added as vectors. (See Figure 3-42.)

Example

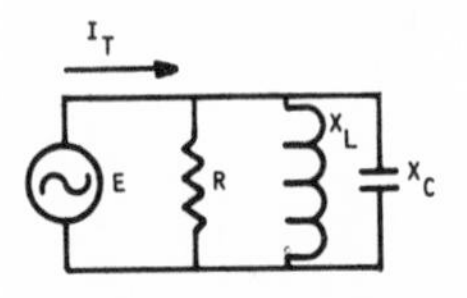

Figure 3-42

Given: $E = 20V$, $R = 20\Omega$, $X_L = 50\Omega$, $X_c = 10\Omega$

Find: I_T, I_R, I_L, I_c

1. $G = \frac{1}{R} = 0.05$ MHos

2. $B_c = \frac{1}{X_c} = 0.10$ MHos

3. $B_L = \frac{1}{X_L} = 0.02$ MHos

$$Y = \text{Admittance}\sqrt{(B_L - B_c)^2 + G^2} = \frac{1}{Z}$$

$$Y = \sqrt{(0.08)^2 + (0.05)^2} = \sqrt{.0089} = 0.09434$$

$$Z = \frac{1}{Y} = 10.6 \text{ ohms}$$

$$I_T = \frac{E}{Z} = \frac{20}{10.6} = 1.887 \text{ amps}$$

$$I_R = \frac{E}{R} = \frac{20}{20} = 1.0 \text{ amps}$$

$$I_L = \frac{E}{X_L} = \frac{20}{50} = 0.40 \text{ amps}$$

$$I_c = \frac{E}{X_c} = \frac{20}{10} = 2.00 \text{ amps}$$

The sum of the Currents $I_r + I_L + I_C$ add to equal I_t, vectorally.

$$I_T = \sqrt{(I_L - I_c)^2 + I_R^2} = \sqrt{1.6^2 + 1^2} = \sqrt{3.56} = 1.887 \text{ amps}$$

The j Operator

It is common to express the impedance of a resistor, capacitor, or inductor circuit using rectangular form or the j operator. (See Figure 3-43.)

Example 1:

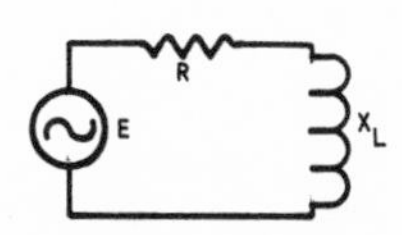

Figure 3-43

Given: $E = 20V$, $R = 50\Omega$, $X_L = 30\Omega$

Find: I

The impedance (Z) can be written as:

$Z = 50 + 30j \longrightarrow j = \sqrt{-1}$

The +30j represents 30 ohms of inductive reactance. Thirty ohms of X_c would be represented as (-30j).

$$I = \frac{E}{Z} = \frac{20}{50 + 30j}$$

Multiply both numerator and denominator by the complex conjugate. (See Chapter 2, Section 7.)

$$I = \frac{20}{50 + 30j} \times \frac{50 - 30j}{50 - 30j} = \frac{1000 - 600j}{2500 - 1500j + 1500j - 900j^2}$$

$$= \frac{1000 - 600j}{2500 - 900j^2} = \frac{10 - 6j}{25 - 9j^2}$$

Since $j^2 = -1$

$$I = \frac{10 - 6j}{25 + 9} = \frac{10 - 6j}{34} = 0.294 - 0.176j$$

This is the rectangular for I:

The magnitude: $I = \sqrt{(0.294)^2 + (0.176)^2} = 0.343$ amps

The current may also be written in polar form.

$I = 0.343 \angle\theta$ Where θ is the phase angle between source voltage and source current

$$\theta = \tan^{-1} \frac{0.176}{0.294} \qquad \theta = 30.9 \text{ degrees}$$

$I = 0.343 \angle 30.9$

Example 2:

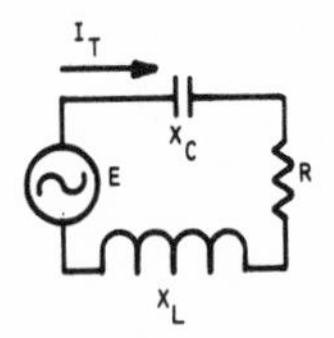

Figure 3-44

Given: $E = 100V$

$X_c = 200\Omega$

$X_L = 50\Omega$

$R = 300\Omega$

Find: I_T, E_c, E_L, E_R

$$I_T = \frac{E}{Z} \qquad Z = 300 - (200 - 50)j$$

$$I_T = \frac{100}{300 - 150j} \qquad I_T = \frac{2}{6 - 3j} \times \frac{6 + 3j}{6 + 3j} = \frac{12 + 6j}{36 - 9j^2}$$

$$I_T = \frac{12 + 6j}{45} = 2.67 + 0.133j \text{ (rectangular form)}$$

$I_T = X \angle\theta$ polar form

Where: $X = \sqrt{2.67^2 + 0.133^2}$

$$\theta = \tan^{-1} \frac{.133}{2.67}$$

$I_T = 2.673 \angle 2.85$

$E_R = I_T R = 2.673 \times 300 = 801.9$ volts

$E_L = I_T X_L = 2.673 \times 50 = 133.65$ volts

$E_c = I_T X_c = 2.673 \times 200 = 534.6$ volts

Series, Parallel Circuits

Circuits consisting of resistors, inductors, and capacitors in a combination of series and parallel can be solved by a combination of the two previous examples. (See Figure 3-45.)

Example:

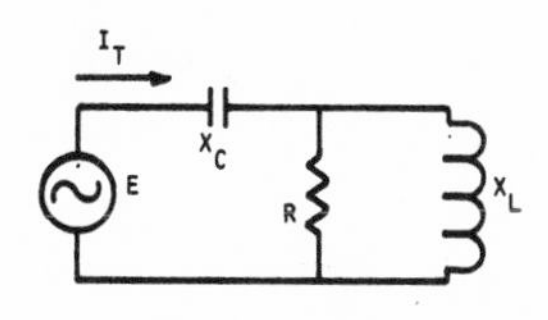

Figure 3-45

Given: $E = 100V$

$X_c = 50\Omega$

$X_L = 100\Omega$

$R = 40\Omega$

Find: I_T

$$I_T = \frac{E}{Z_T} \quad Z_T \text{ must be determined}$$

$$G = \frac{1}{R} = 0.025$$

$$B_L = \frac{1}{X_L} = \frac{1}{100j} = -0.01j$$

$$Y = 0.025 - 0.01j$$

$$Z = \frac{1}{Y} = \frac{1}{.025 - .01j}$$

$$Z = \frac{1}{.025 - .01j} \times \frac{.025 + .01j}{.025 + .01j} = \frac{.025 + .01j}{.000625 - .0001j^2}$$

$$Z = 34.48 + 13.79j$$

$$Z_T = 34.48 + 13.79j - 50j$$

$$Z_T = 34.48 - 36.21j$$

$$I_T = \frac{E}{Z_T} = \frac{100}{34.48 - 36.21j}$$

$$I_T = \frac{100}{34.48 - 36.21j} \times \frac{34.48 + 36.21j}{34.48 + 36.21j} = \frac{3448 + 3621j}{1188.87 + 1311.16}$$

$$I_T = 1.379 + 1.448j \text{ (rectangular form)}$$

$$I_T \; 1.9999 \angle 46.4^\circ \text{ polar}$$

HOW TO DETERMINE RESONANCE

Resonance occurs in an A.C. circuit whenever inductive reactance equals capacitive reactance.

$$X_L = X_C$$

$$2\pi fL = \frac{1}{2\pi fC}$$

$$4\pi^2 f^2 LC = 1$$

$$f^2 = \frac{1}{4\pi^2 LC}$$

$$f = \frac{1}{2\pi\sqrt{LC}}$$

In a series resonant circuit, the inductive and capacitive reactance cancel, so that the resistance is the only impedance left. A pure L C series resonance circuit is a short circuit. (See Figure 3-46.)

Example:

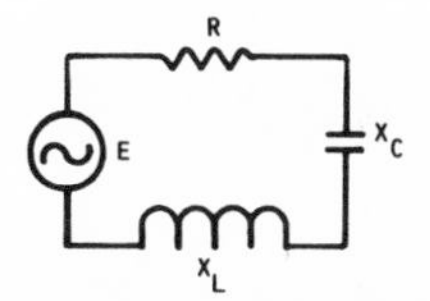

Figure 3-46

Given: $E = 25V$

$R = 5\Omega$

$X_L = 1000\Omega$

$X_C = 1000\Omega$

Find: I, E_L, E_C

$$I = \frac{E}{Z} \qquad Z = \sqrt{(X_L - X_C)^2 + R^2} = R$$

$$I = \frac{25}{5} = 5 \text{ amps}$$

$$E_C = IX_C = 5 \times 1000 = 5000 \text{ volts}$$

$$E_L = IX_L = 5 \times 1000 = 5000 \text{ volts}$$

In a parallel resonant circuit, the inductive and capacitive reactance cancel, but the impedance is infinite. A pure parallel resonant circuit is an open circuit. (See Figure 3-47.)

Example:

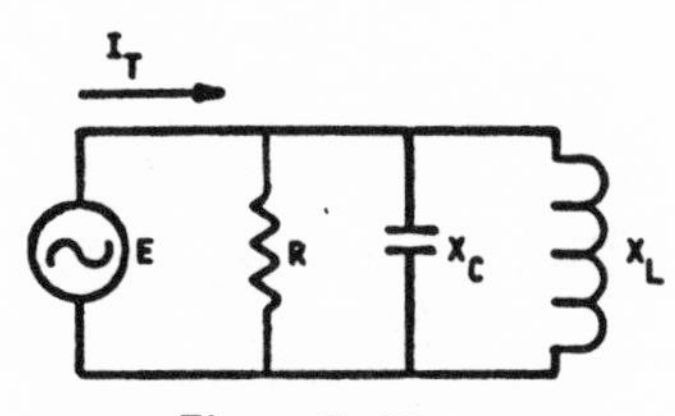

Figure 3-47

Given: $E = 80V$, $R = 80\Omega$, $X_L = 1\Omega$, $X_C = 1\Omega$

Find: I_T

$$I_T = \sqrt{I_R^2 + (I_L - I_C)^2}$$

$$I_R = \frac{E}{R} = \frac{80}{80} = 1.0 \text{ amp}$$

$$I_L = \frac{E}{X_L} = \frac{80}{1} = 80 \text{ amps}$$

$$I_C = \frac{E}{X_C} = \frac{80}{1} = 80 \text{ amps}$$

$$I_T = \sqrt{1^2 + 0^2} = 1.0 \text{ amp}$$

All the current from the source (1.0 amp) goes through the resistor.

Boolean Algebra is the mathematical theory on which all modern computers operate, from the large scientific computers to the small logic boards designed for a single purpose. A working knowledge of its concepts is essential for anyone working in electronics.

This chapter clearly outlines the basic concepts and gives worked-out examples wherever possible. The chapter starts with the three basic gates or connectives used in Boolean Algebra and defines their input/output relationships. Since all logic circuits are made from these three basic functions, the input/output relationships for any logic circuit can be determined from knowledge of AND gates, OR gates, and inverters.

There are three ways of expressing a Boolean Algebra function: (1) a Truth Table, (2) a Switching Function, and (3) a Logic Diagram. Given any one of the three, the other two can be derived. This chapter gives the methods, both graphical and algebraic, for deriving the desired form. There are practical problems at the end of the chapter that show, step by step, how to derive a logic circuit to solve a given problem.

In Boolean Algebra, it is fundamental that a variable may have only two values. These values are usually represented by zero (0) or one (1). They may represent a switch that is on or off, a relay that is energized or de-energized, or they may represent a voltage difference.

+5 volts = 1

0 volts = 0

When applied to an equation such as W = A + B, the values of W, A, and B may either be one or zero, instead of the entire range of real numbers they could represent in other algebras.

SECTION 1: THE THREE BASIC GATES

Boolean Algebra consists of three basic connectives: logical addition (the OR gate), logical multiplication (the AND gate), and negation (the inverter). All switching functions represent some combination of these three basic functions.

The OR Gate

The OR gate is most commonly drawn as in Figure 4-1D. It can also be drawn with these other symbols.

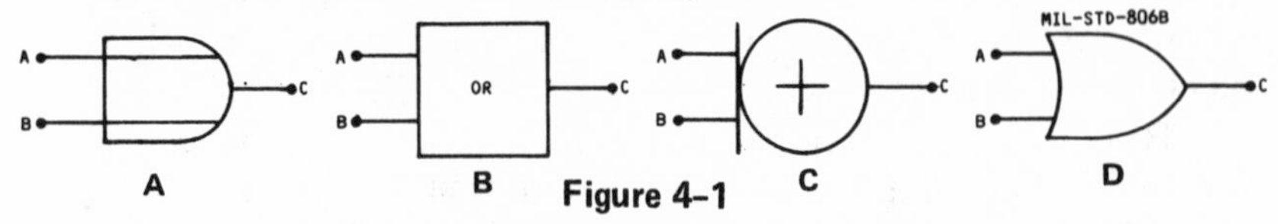

Figure 4-1

The input/output relationship is shown in Figure 4-2. The output "C" is 1 when either of the inputs (A or B) is a 1. The table in Figure 4-2 is defined as logical addition or addition in Boolean Algebra. It is written as follows: A + B = C. The "+" sign implies that the variables A and B are being applied to an OR gate.

A	B	C
0	0	0
0	1	1
1	0	1
1	1	1

Figure 4-2

The AND Gate

The AND gate is most commonly drawn as in Figure 4-3C. It can also be drawn with these other symbols.

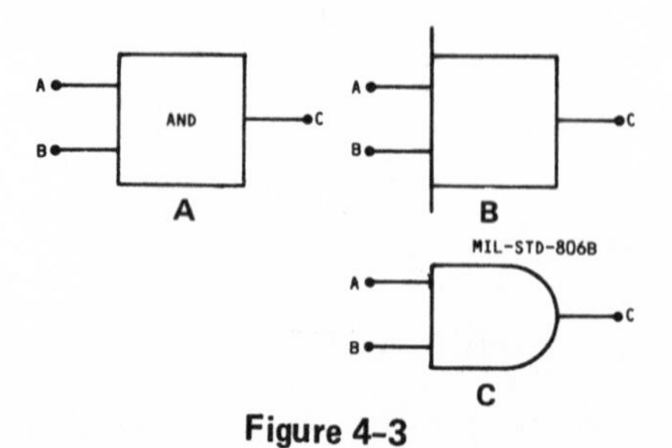

Figure 4-3

The input/output relationship is shown in Figure 4-4. The output "C" is 1 only when both (A and B) are ones. The table in Figure 4-4 is defined as logical multiplication or multiplication in Boolean Algebra. It is written as follows: AB = C

A	B	C
0	0	0
0	1	0
1	0	0
1	1	1

Figure 4-4

The multiplication (A X B) implies that the variables A and B are being applied to an AND gate.

The Inverter (Negation)

The Inverter is most commonly drawn as follows:

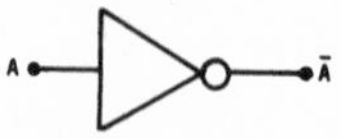

A —▷○— $\bar{A}$

Figure 4-5

It has the input/output relationship shown in Figure 4-6. The output is the inversion (negation) of the input.

A	$\bar{A}$
0	1
1	0

Figure 4-6

Inverted Functions (NANDS, NORS)

It is common to use inverters on the output of AND/OR gates, so these gates are given their own names. When the output of a gate is inverted, it is referred to as follows in Figure 4-7. The small circle on the output means the signal has been inverted or negated. The circle means the same if used on the input. (See Figure 4-8.)

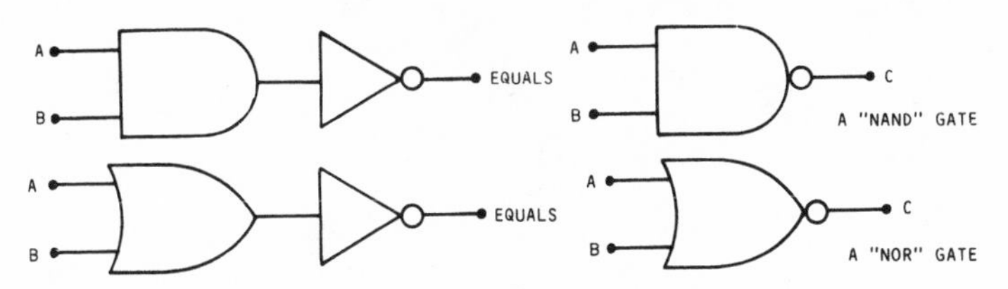

Figure 4-7

Example:

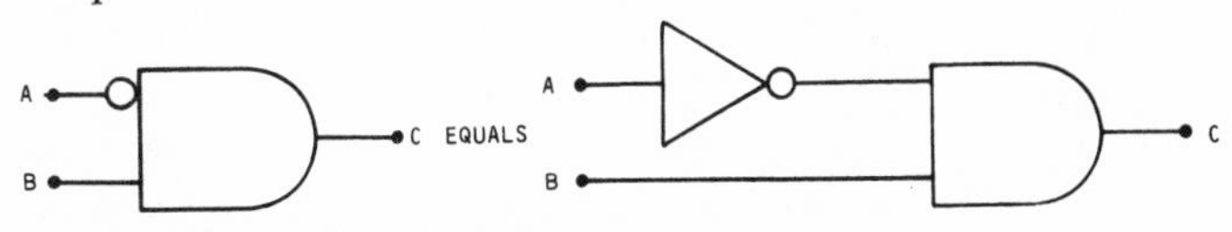

Figure 4-8

Multiple Inputs

Gates may have more than the two inputs used in the aforementioned examples. With multiple inputs, the OR gate will have an output "1" whenever any input is "1." The AND gate must have all inputs as "1" before the output is "1." (See Figure 4-9A and B.)

Examples:

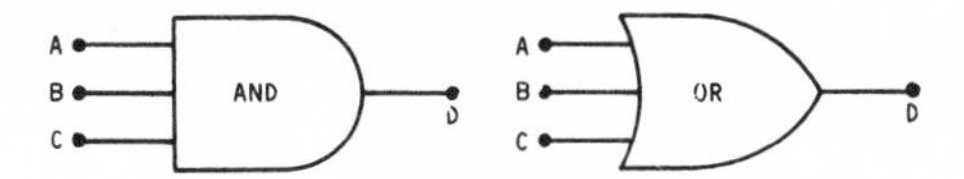

Figure 4-9A

A	B	C	D
0	0	0	0
0	0	1	0
0	1	0	0
0	1	1	0
1	0	0	0
1	0	1	0
1	1	0	0
1	1	1	1

A	B	C	D
0	0	0	0
0	0	1	1
0	1	0	1
0	1	1	1
1	0	0	1
1	0	1	1
1	1	0	1
1	1	1	1

Figure 4-9B

SECTION 2: HOW TO GENERATE SWITCHING FUNCTIONS FROM TRUTH TABLES

There are three basic ways of expressing a Boolean Algebra function: a Truth Table, a Switching Function, and a Logic Diagram. This section is concerned with defining these three expressions, along with methods used to convert between them.

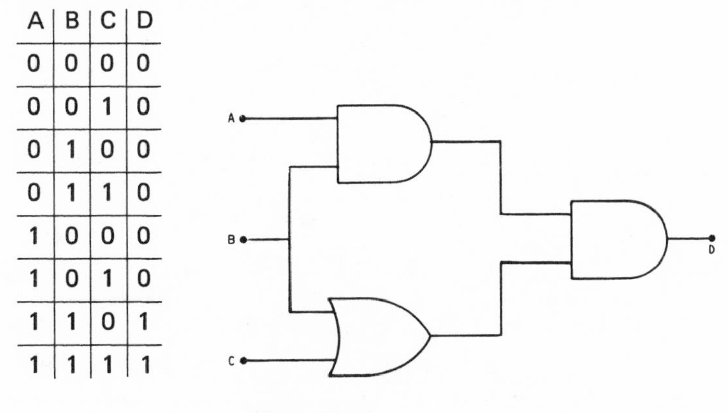

A	B	C	D
0	0	0	0
0	0	1	0
0	1	0	0
0	1	1	0
1	0	0	0
1	0	1	0
1	1	0	1
1	1	1	1

Figure 4-10

A Truth Table is a map showing the output of a logic circuit for each possible input.

The logic diagram of Figure 4-10 is completely defined by the Truth Table. The inputs are A, B, and C, and all possible combinations of A, B, and C are listed. (There are eight in this case.) The output D will be one, with only two combinations of A, B, and C.

Combinations of Inputs

If there are two inputs, the only combinations of these two are found in Figure 4-11.

Four Total Combinations

1.	A = 0	B = 0
2.	A = 0	B = 1
3.	A = 1	B = 0
4.	A = 1	B = 1

Figure 4-11

For three inputs: (See Figure 4-12.)

Eight Total Combinations

1.	A = 0	B = 0	C = 0
2.	A = 0	B = 0	C = 1
3.	A = 0	B = 1	C = 0
4.	A = 0	B = 1	C = 1
5.	A = 1	B = 0	C = 0
6.	A = 1	B = 0	C = 1
7.	A = 1	B = 1	C = 0
8.	A = 1	B = 1	C = 1

Figure 4-12

The simplest way to list all the combinations is to count in binary from 000 to 111, top to bottom. In general, the number of input combinations is 2^n, where (N) is the number of inputs.

Combinations of Outputs

Consider the two input Truth Tables where the two inputs are A and B. One possible output is C, all zeros; another is C_1, and another is C_2. The general rule says that the total number of output combinations is seen in Figure 4-13.

A	B	C	C_1	C_2	C_3	C_4	C_5	C_6
0	0	0	0	0	0	0	0	0
0	1	0	0	0	0	1	1	1
1	0	0	0	1	1	0	0	1
1	1	0	1	0	1	0	1	0

Figure 4-13

$$2^{(2)^{(N)}}$$

Example: A four-input circuit has 65,536 different output possibilities.

$2^N = 2^4 = 16, \quad 2^{16} = 65{,}536$

Switching Functions

The switching functions is a Boolean Algebra representation of a logic circuit. Consider the circuit in Figure 4-14.

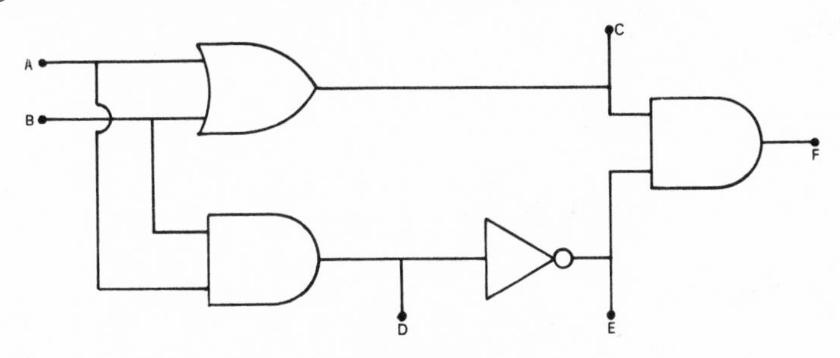

Figure 4-14

A. The output at C is C = A + B.
B. The output at D is D = AB.

C. The output at E is $E = \overline{AB}$.
D. The output at F is $F = CE$.

$$F = (A + B)\ AB$$

The equation $F = (A + B)\ \overline{AB}$ is a switching function for the circuit in Figure 4-14.

Sum of Products

Sum-of-Products is one way in which a switching function can be derived directly from a Truth Table. For example, consider the Truth Table in Figure 4-15. To generate a switching function by Sum-of-Products, consider only the outputs that are one. They are:

$A = 0 \quad B = 1 \quad C = 1$
$A = 1 \quad B = 0 \quad C = 1$

A	B	C
0	0	0
0	1	1
1	0	1
1	1	0

1's output
0's output

Figure 4-15

The two products that give 1 on the output are:

$\overline{A}B \qquad A\overline{B}$

The sum of these products give:

$\overline{A}B + A\overline{B} = C$

This is a switching function for this Truth Table. The logic diagram representing this switching function looks like Figure 4-16. It is an OR gate fed by AND gates.

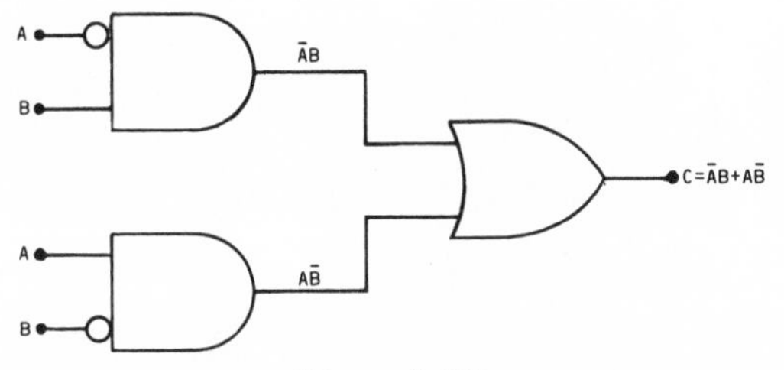

Figure 4–16

Product-of-Sums

Again consider the Truth Table in Figure 4–15. Pick the rows where the output C is zero. They are:

$A = 0 \quad B = 0 \quad C = 0$
$A = 1 \quad B = 1 \quad C = 0$

Where the variables equal zero, write them as A, B. Where they are ones, write them as $\bar{A}$, $\bar{B}$. The two sums are:

$$(A + B) \qquad (\bar{A} + \bar{B})$$

Their product is:

$$(A + B)(\bar{A} + \bar{B})$$

The switching function is

$$C = (A + B)(\bar{A} + \bar{B})$$

The logic diagram for this switching function is seen in Figure 4–17. It is an AND gate fed by OR gates.

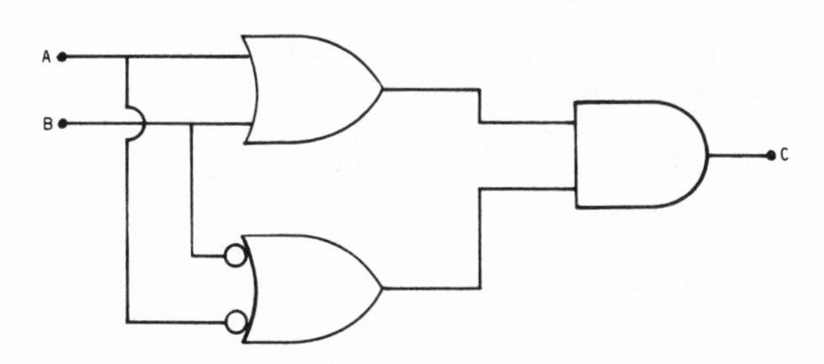

Figure 4–17

Example:

Consider the two input Truth Tables in Figure 4–18. To generate the switching function by Sum-of-Products:

A	B	C	
0	0	1	–Output 1
0	1	0	–Output 2
1	0	0	–Output 3
1	1	1	–Output 4

Figure 4–18

1. Determine the 1 outputs. They are:
   A = 0 B = 0 and A = 1 B = 1
2. Write down the products of these two rows. They are:
   $\overline{A}\overline{B}$ and AB
3. Add the two Products:
   $C = \overline{A}\overline{B} + AB$

The logic diagram for this switching function is:

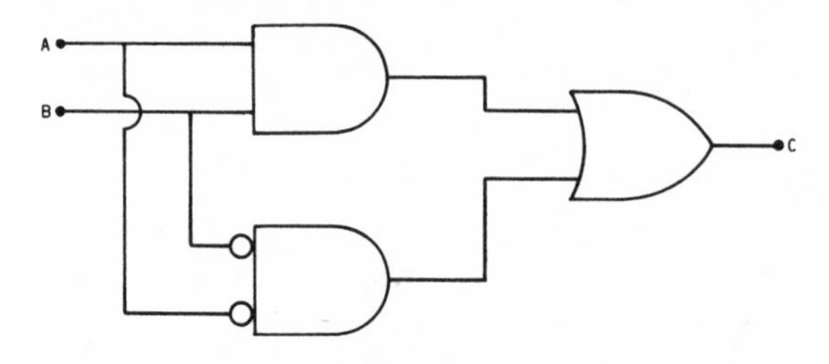

Figure 4–19

An OR gate fed by two AND gates.
To generate a switching function by Product-of-Sums:

1. Determine the 0 outputs. They are:
   A = 0 B = 1 and A = 1 B = 0
2. Invert the variable: If A = 0 Write A, if A = 1 write $\overline{A}$.
3. The two sums are:
   $A + \overline{B}$ and $\overline{A} + B$

4. Multiply the two sums:

   $C = (A + \overline{B})(\overline{A} + B)$

The logic diagram for this switching function is:

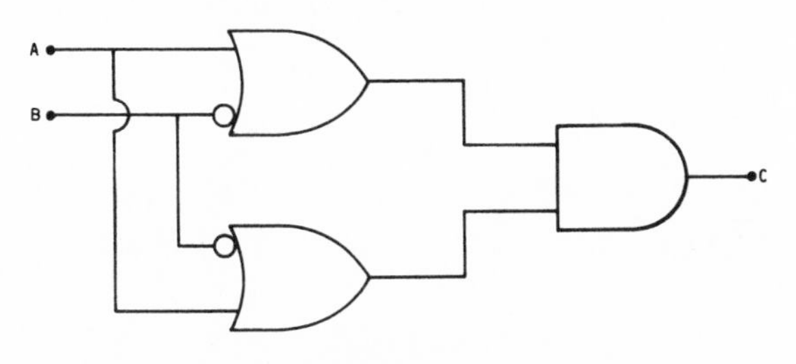

Figure 4-20

An AND gate fed by OR gates.

It will be shown in the following sections that the two switching functions and logic diagrams are equal.

SECTION 3: CHECKLIST OF BOOLEAN ALGEBRA RULES

The rules of Boolean Algebra are useful relations that allow the simplification or manipulation of switching functions. To understand the relationship, apply the rule to a gate network.

Example:

1. $A \cdot \overline{A} = 0$
2. The $\cdot$ represents an AND gate.
3. A, $\overline{A}$ (inputs to AND gate)
4. For a 1 output, both inputs must be 1. Since the inputs are A and $\overline{A}$, both cannot be 1 at the same time.

Boolean Algebra Rules

1. $A \cdot 0 = 0$
2. $A \cdot 1 = A$
3. $A \cdot A = A$
4. $A \cdot \overline{A} = 0$
5. $\overline{\overline{A}} = A$
6. $A + 0 = A$

Boolean Algebra Rules

7. $A + 1 = 1$
8. $A + A = A$
9. $A + \bar{A} = 1$
10. $A + B = B + A$
11. $AB = BA$
12. $A + (B + C) = (A + B) + C$
13. $A(BC) = (AB)C$
14. $A(B + C) = AB + AC$
15. $A + AB = A$
16. $A(A + B) = A$
17. $(A + B)(A + C) = A + BC$
18. $A + \bar{A}B = A + B$

A. Rules 10 and 11 are known as the *Commutative Laws.*

B. Rules 12 and 13 are known as the *Associative Laws.*

C. Rule 14 is the *Distributive Law.*

DeMorgan's Theorem

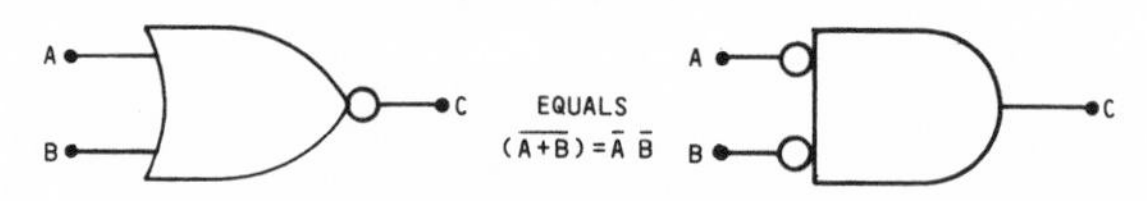

Figure 4–21

DeMorgan's Theorem says that you can change the inversion from the output to the input (or vice versa). You can change the function AND to OR or OR to AND. (See Figures 4–21 and 4–22.)

Figure 4–22

Positive vs. Negative Logic

It is quite common in digital logic to use positive 5 volt logic. That is, +5 volts equals 1 and 0 volts equals 0. It is also possible to reverse the definition and let 0 volts equal 1, and +5 volts equal 0. In general, any voltage or current levels can be used to represent ones and zeros.

Definitions:

Positive Logic—In use when the "more positive" voltage level is used to represent ones and the "more negative" voltage level represents zeros. An example of positive logic is: +5 Volts = 1, 0 volts = 0. Another is -5 volts = 1, -10 volts = 0.

Negative Logic—In use when the "more negative" voltage level is used to represent ones and the "more positive" voltage level represents zeros. An example of negative logic is: +5 volts = 1, +10 volts = 0.

When the definition is changed from positive logic to negative logic (or vice versa), the basic gates are changed. AND gates become OR gates and OR gates become AND gates. (See Figure 4-23.)

Example:

A positive logic AND gate.

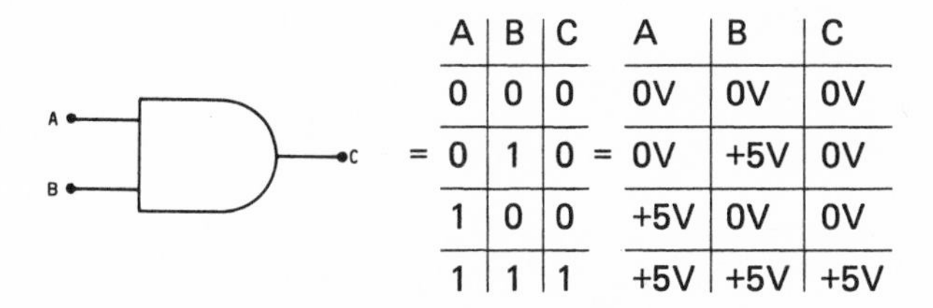

A	B	C
0	0	0
0	1	0
1	0	0
1	1	1

A	B	C
0V	0V	0V
0V	+5V	0V
+5V	0V	0V
+5V	+5V	+5V

Figure 4-23

If the definition is changed to negative logic, the Truth Table becomes Figure 4-24.

A	B	C
1	1	1
1	0	1
0	1	1
0	0	0

=

Figure 4-24

SECTION 4: HOW TO SIMPLIFY AND MANIPULATE EQUATIONS

Using the rules of Section 3, switching functions can be manipulated so the gates used are those desired, or simplified so that the number of gates used is a minimum. The following equations (Figures 4-25 and 4-26) show how this is done.

Example 1: Change the switching function to a different form.

$C = (A + \overline{B})(\overline{A} + B)$

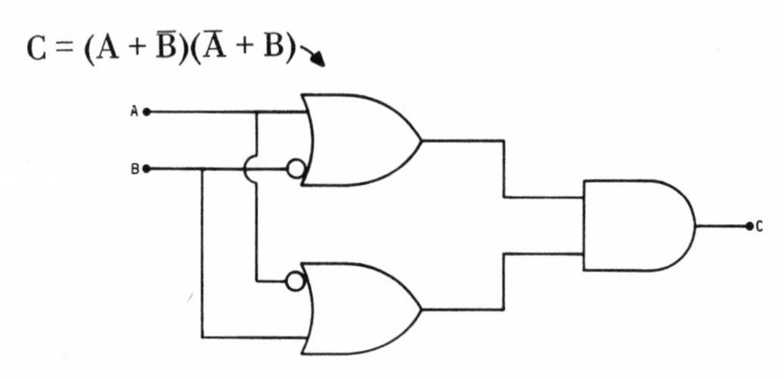

Figure 4-25

Multiply the two sums:

$$\begin{array}{l} A + \overline{B} \\ \times \overline{A} + B \\ \hline \overline{A}A + \overline{A}\overline{B} + AB + \overline{B}B \end{array}$$

$\overline{A}A = B\overline{B} = 0$ Rule 4

So: $C = \overline{A}\overline{B} + AB$

Figure 4-26

The Truth Table is the same for both switching functions.

Example 2: Draw a logic diagram out of NAND gates for the switching function, as seen in Figures 4-27 and 4-28.

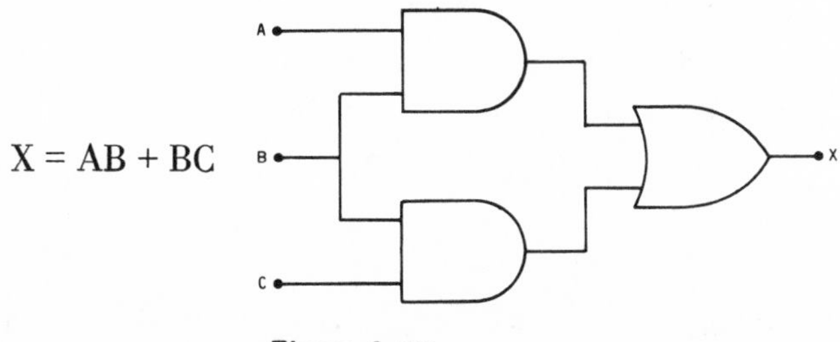

Figure 4-27

Using Rule 5: $X = AB + BC = \overline{\overline{AB + BC}}$

DeMorgan $X = \overline{\overline{AB + BC}} = \overline{\overline{AB} \cdot \overline{BC}}$

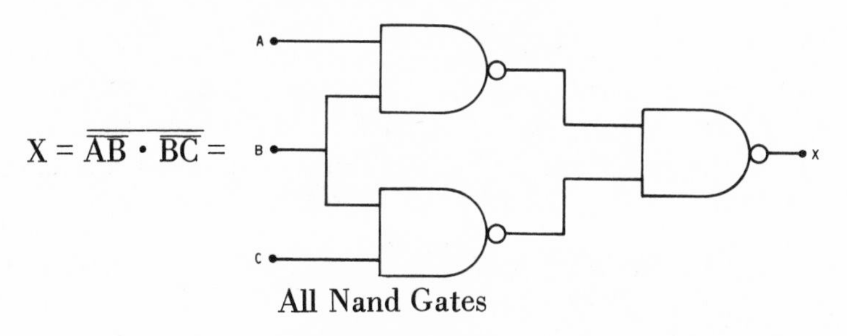

All Nand Gates

Figure 4-28

Example 3: Simplify the following function:

$A = XY + XYZ + XY\overline{Z} + \overline{X}YZ$

$A = (XY + XYZ) + XY\overline{Z} + \overline{X}YZ$

$A = XY(1 + Z) + XY\overline{Z} + \overline{X}YZ$

$A = XY + XY\overline{Z} + \overline{X}YZ \qquad (1 + Z) = 1$

$A = (XY + XY\overline{Z}) + \overline{X}YZ$

$A = XY(1 + \overline{Z}) + \overline{X}YZ$

$A = XY + \overline{X}YZ \qquad (1 + \overline{Z}) = 1$

$A = Y(X + \overline{X}Z)$

$(X + \overline{X}Z) = X + Z \qquad$ Rule 18

$A = YX + YZ$

Example 4: Simplify the following function:

$$X = ABC(AB\overline{C} + A\overline{B}C + \overline{A}BC)$$
$$X = (ABC)(AB\overline{C}) = 0$$
$$+ (ABC)(A\overline{B}C) = 0$$
$$+ (ABC)(\overline{A}BC) = 0$$
$$X = 0$$

This function cannot have an output of 1. Note that $ABC \cdot AB\overline{C}$ equals zero since $C\overline{C} = 0$ and $0 \cdot AB = 0$.

Example 5: Simply the following function:

$$X = AB + A\overline{B} + \overline{A}C + \overline{A}\overline{C}$$

Look for terms where one and only one term changes such as: $\overline{A}C$ and $\overline{A}\overline{C}$. C changed from C to $\overline{C}$. Combine these terms and factor:

$$X = (AB + A\overline{B}) + (\overline{A}C + \overline{A}\overline{C})$$
$$X = A(B + \overline{B}) + \overline{A}(C + \overline{C})$$
$$X = A + \overline{A}$$
$$X = 1$$

Example 6: Given the Truth Table, draw a logic diagram to fit, using a minimum number of gates, as in Figure 4–29 and 4–30.

A	B	C
0	0	1
0	1	1
1	0	0
1	1	0

Figure 4–29

A. Using Sum-of-Products

$C = \overline{A}\overline{B} + \overline{A}B$

$C = \overline{A}(\overline{B} + B)$

$C = \overline{A}$ =

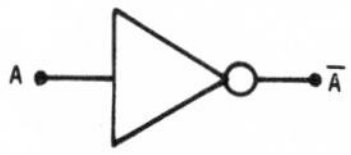

Figure 4-30

Example 7: Given the Truth Table, draw a logic diagram to fit, using a minimum number of gates. (See Figure 4-31.)

A	B	C	D
0	0	0	1
0	0	1	1
0	1	0	1
0	1	1	0*
1	0	0	1
1	0	1	0*
1	1	0	1
1	1	1	1

Figure 4-31

A. Use product-of-sums, since there are fewer zeros in the output than there are ones.

$D = (A + \overline{B} + \overline{C})(\overline{A} + B + \overline{C})$

This switching function uses three gates, as seen in Figure 4-32.

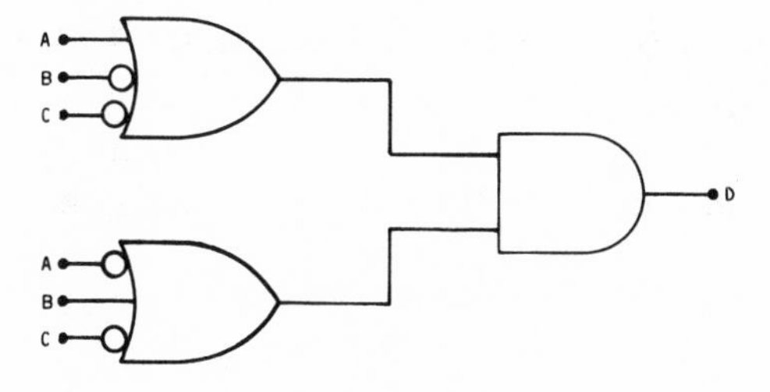

Figure 4-32

Multiplying the two sums:

$$
\begin{array}{r}
A + \overline{B} + \overline{C} \\
\times \overline{A} + B + \overline{C} \\
\hline
\end{array}
$$

$$\overline{A}A + \overline{A}\overline{B} + \overline{A}\overline{C} + AB + B\overline{B} + B\overline{C} + A\overline{C} + \overline{B}\overline{C} + \overline{C}\overline{C}$$

$$\overline{A}A = B\overline{B} = 0, \overline{C}\overline{C} = \overline{C}$$

$$D = \overline{A}\overline{B} + \overline{A}\overline{C} + AB + B\overline{C} + A\overline{C} + \overline{B}\overline{C} + \overline{C}$$

Factor $\overline{C}$ from each possible product:

$$D = \overline{C}\underbrace{(\overline{A} + B + A + \overline{B} + 1)} + \overline{A}\overline{B} + AB$$

This is an OR gate, with a one on the last input, therefore the output is one.

$$D = \overline{C} + \overline{A}\overline{B} + AB$$

Figure 4-33

This circuit is not simpler than the original circuit. It is a different arrangement. (See Figure 4-33.)

SECTION 5: USING MAPS TO SIMPLIFY

Switching functions may be manipulated and simplified using algebra and the list of rules in Section 3. In addition, there are graphical methods that are used to determine if a switching function can be simplified. One of these is the N-cube.

If there is one variable "A," this variable can have two values 0 and 1. This can be represented by a line segment. (See Figure 4-34.)

Figure 4-34

The above example cannot be simplified since the dots are on opposite vertices.

How to Simplify

Example:

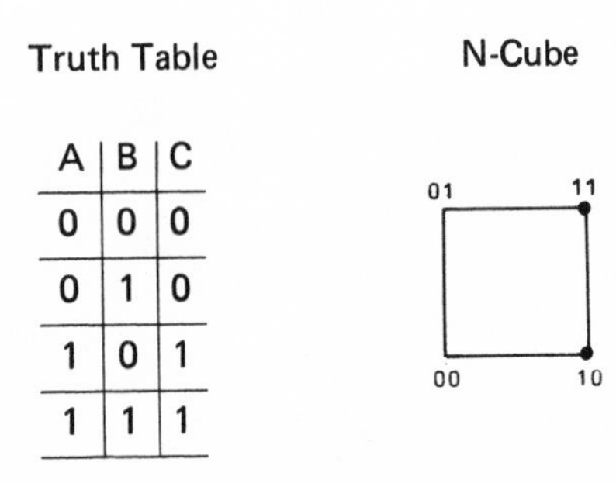

A	B	C
0	0	0
0	1	0
1	0	1
1	1	1

Figure 4-37

Sum-of-Products = $C = A\overline{B} + AB$

On the N-Cube in Figure 4-37, adjacent vertices are dotted, $(A = 1)(B = 0)$, $(A = 1)(B = 1)$. When this occurs, the two terms are combined and the variable that changed along the line is eliminated. In this case B changed, so the function becomes:

$$C = A$$

Note the sum-of-products function:

$$C = A\overline{B} + AB$$
$$C = A(\overline{B} + B)$$
$$C = A$$

So the same result can be obtained through algebra. (See Figure 3-48.)

Going one more step, two variables can be represer square. There are four possible combinations of A an these combinations are represented by the vertices of th This square can quickly show if a given function can be si (See Figure 4–35.)

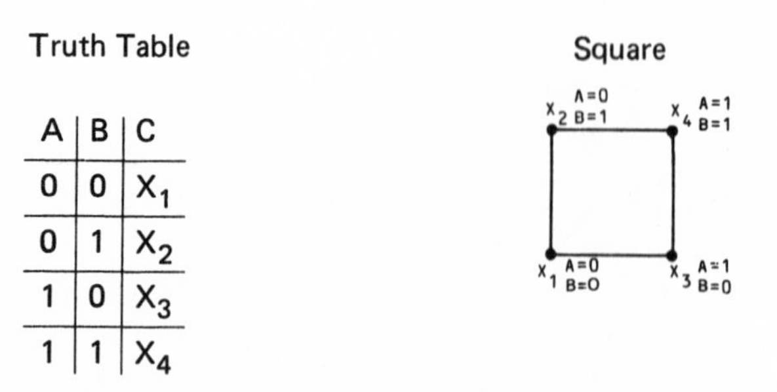

Truth Table

A	B	C
0	0	X_1
0	1	X_2
1	0	X_3
1	1	X_4

Figure 4–35

Example:

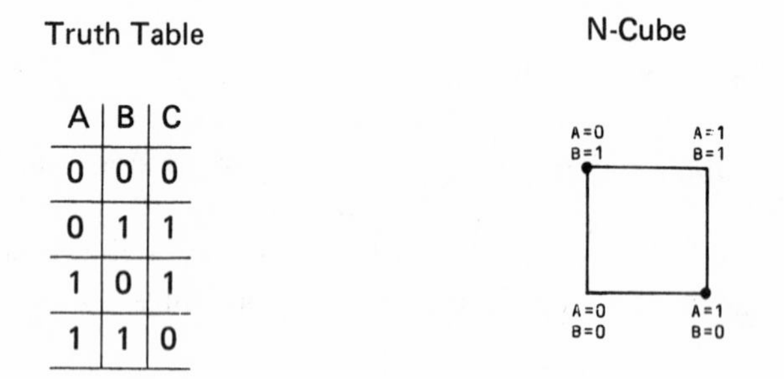

Truth Table

A	B	C
0	0	0
0	1	1
1	0	1
1	1	0

Figure 4–36

$C = \overline{A}B + A\overline{B}$ Sum-of-Products

The "Dots" on the vertices represent the ones out Truth Table. (See Figure 4–36.)

Rule:

If adjacent vertices on the N-Cube have dots, the fi be simplified. If there are no adjacent dots, the functio simplified.

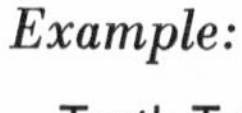

Example:

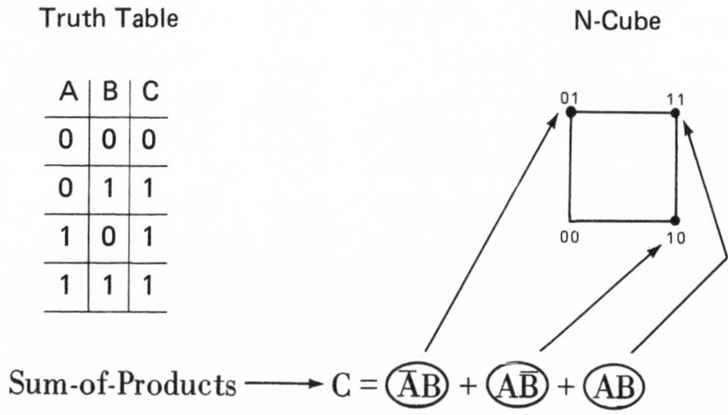

A	B	C
0	0	0
0	1	1
1	0	1
1	1	1

Sum-of-Products ⟶ $C = \bar{A}B + A\bar{B} + AB$

Figure 4–38

Across the top line (the vertice representing $\bar{A}B$ and AB) the A changes, so the two terms are combined and A is dropped, which leaves:

$C = B +$ (terms from vertical line)

Along the right vertical line, B changes so it can be dropped from the terms that represent the two vertices.

$C = B + A$

A simple OR Gate.

The Three-Input Table

A one-variable is represented by a line segment, a two-variable by a square, and a three-variable by a cube. (See Figure 4–39.)

Example:

Truth Table

A	B	C	D
0	0	0	X_1
0	0	1	X_2
0	1	0	X_3
0	1	1	X_4
1	0	0	X_5
1	0	1	X_6
1	1	0	X_7
1	1	1	X_8

N-Cube

X_4 011
X_3 010
111 X_8
110 X_7
X_2 001
101 X_6
000 X_1
100 X_5

Figure 4–39

The same rule applies that, if adjacent vertices have dots, the sum-of-product can be simplified and the variable that changes along that line can be eliminated.

Example:

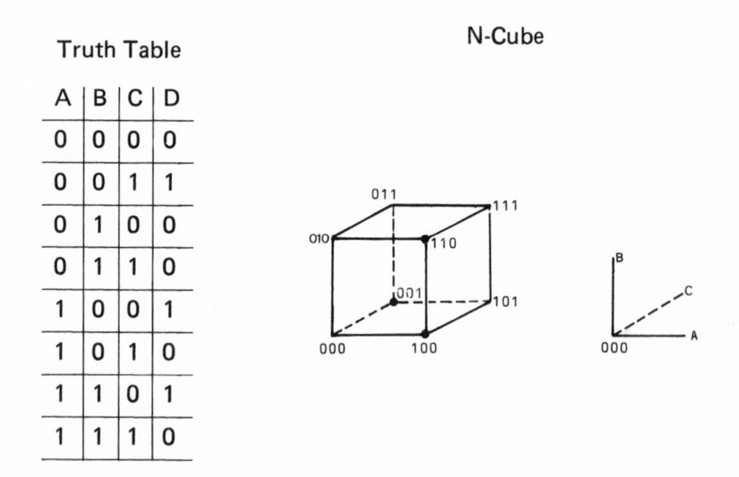

Truth Table

A	B	C	D
0	0	0	0
0	0	1	1
0	1	0	0
0	1	1	0
1	0	0	1
1	0	1	0
1	1	0	1
1	1	1	0

Figure 4–40

$$D = \bar{A}\bar{B}C + A\bar{B}\bar{C} + AB\bar{C}$$

In Figure 4-40, the only adjacent vertices are 100, 110. Along this line B changes; it can be dropped and the two terms can be combined.

$$D = \bar{A}\bar{B}C + A\bar{C}$$

From algebra:

$$D = \bar{A}\bar{B}C + A\bar{B}\bar{C} + AB\bar{C}$$

$$D = \bar{A}\bar{B}C + A\bar{C}(\bar{B} + B)$$

$$\bar{B} + B = 1$$

$$D = \bar{A}\bar{B}C + A\bar{C}$$

Example:

Truth Table

A	B	C	D
0	0	0	0
0	0	1	0
0	1	0	1
0	1	1	1
1	0	0	0
1	0	1	0
1	1	0	1
1	1	1	1

N-Cube

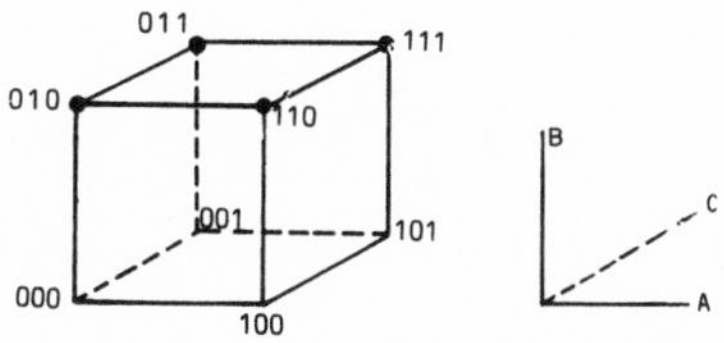

Figure 4-41

In Figure 4-41, all the dots are on adjacent vertices. The only variable that didn't change was B, so

$$D = B$$

The Four Input Table

Four input Truth Tables can be handled by the N-Cube method, but it becomes difficult. The cube becomes a "Hyper" cube or four-dimensional cube. (See Figure 4–42.)

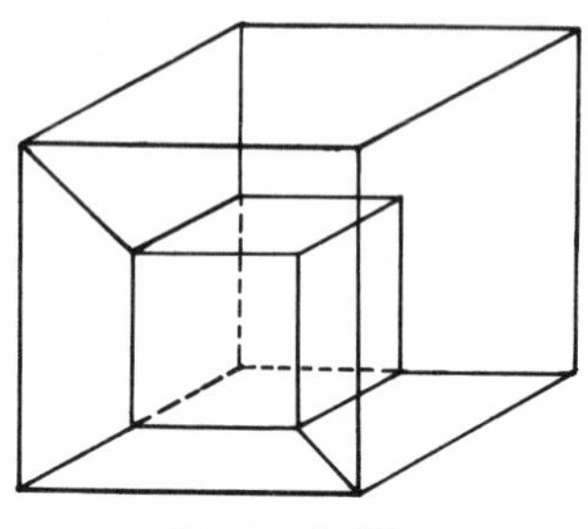

Figure 4–42

The Karnaugh Map

The Karnaugh map is another graphical method by which a switching function can be simplified. Similar to the N-Cube, the Karnaugh Map is a way to represent a Truth Table so that switching functions can be derived in their simplest form. (See Figure 4-4-43.)

The Two-Variable Map

Truth Table

A	B	C
0	0	X_1
0	1	X_2
1	0	X_3
1	1	X_4

Karnaugh Map

B \ A	$\overline{A}$	A
$\overline{B}$	X_1	X_3
B	X_2	X_4

or

B \ A	0	1
0	X_1	X_3
1	X_2	X_4

Figure 4–43

Three Variable Map

Truth Table

A	B	C	D
0	0	0	X_1
0	0	1	X_2
0	1	0	X_3
0	1	1	X_4
1	0	0	X_5
1	0	1	X_6
1	1	0	X_7
1	1	1	X_8

Karnaugh Map

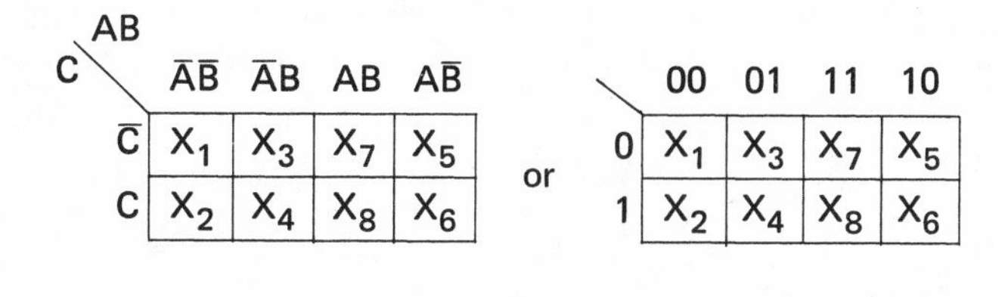

Figure 4-44

Four Variable Map

Truth Table

A	B	C	D	Z
0	0	0	0	X_1
0	0	0	1	X_2
0	0	1	0	X_3
0	0	1	1	X_4
0	1	0	0	X_5
0	1	0	1	X_6
0	1	1	0	X_7
0	1	1	1	X_8
1	0	0	0	X_9
1	0	0	1	X_{10}
1	0	1	0	X_{11}
1	0	1	1	X_{12}
1	1	0	0	X_{13}
1	1	0	1	X_{14}
1	1	1	0	X_{15}
1	1	1	1	X_{16}

Karnaugh Map

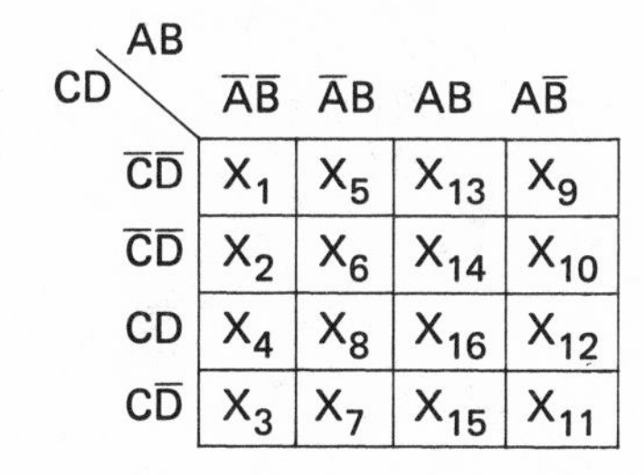

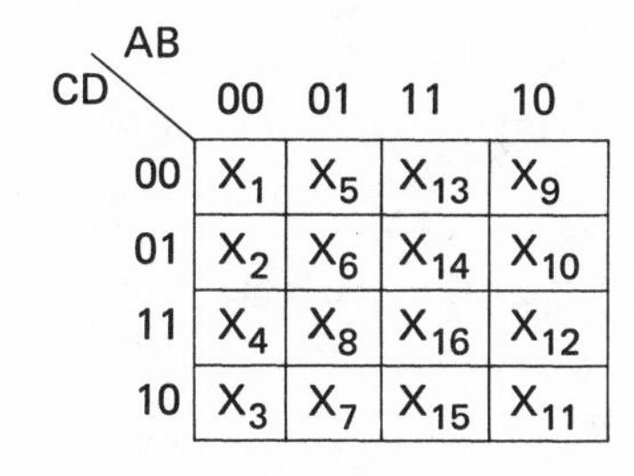

Figure 4-45

Note 1: The vertical and horizontal scales are written in "Gray" code. One and only one variable changes in subsequent steps. (See Figures 4-44 and 4-45.)

Note 2: Karnaugh Maps are considered to be "rolled." The top and bottom lines are considered as touching so that X_3, X_7, X_{15}, and X_{11} are adjacent to X_1, X_5, X_{13}, and X_9. This is also true for the left and right hand lines, so that X_9, X_{10}, X_{12}, and X_{11} are adjacent to X_1, X_2, X_4, and X_3.

Simplifying

Rule 1: If a one appears in adjacent cells, the function can be simplified. (See Figure 4-46.)

Example:

Truth Table

A	B	C
0	0	1
0	1	1
1	0	0
1	1	0

Karnaugh Map

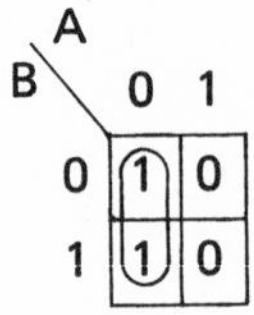

B \ A	0	1
0	1	0
1	1	0

Figure 4-46

$$C = \overline{A}\overline{B} + \overline{A}B$$

"B" is the variable that changes between the two adjacent cells, therefore the two terms can be combined and "B" can be eliminated:

$$C = \overline{A}$$

Example:

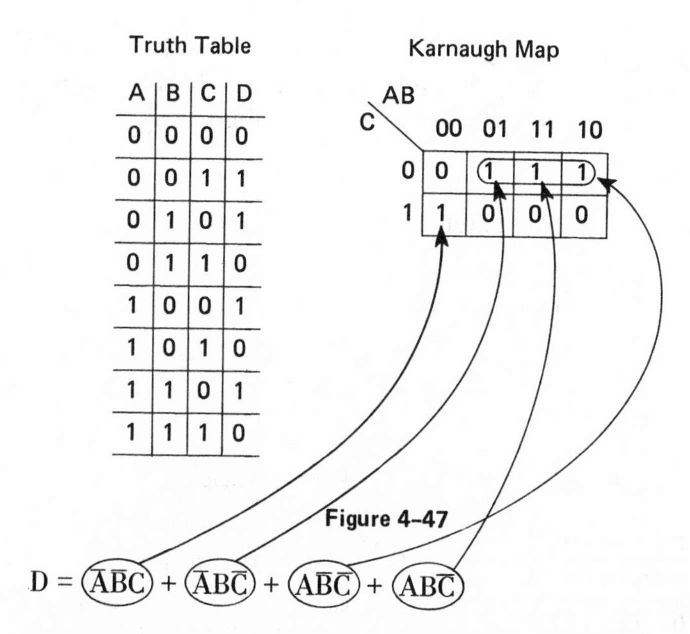

Truth Table

A	B	C	D
0	0	0	0
0	0	1	1
0	1	0	1
0	1	1	0
1	0	0	1
1	0	1	0
1	1	0	1
1	1	1	0

Figure 4-47

$$D = \overline{A}\,\overline{B}C + \overline{A}B\overline{C} + A\overline{B}\,\overline{C} + AB\overline{C}$$

In the map, there are three adjacent ones. The only variable that did not change in this group is C. (See Figure 4-47.) Therefore:

$$D = \overline{A}\,\overline{B}C + \overline{C}$$

The first two examples were straightforward, since there were no intersecting adjacent cubes. Consider the following:

A	B	C
0	0	0
0	1	1
1	0	1
1	1	1

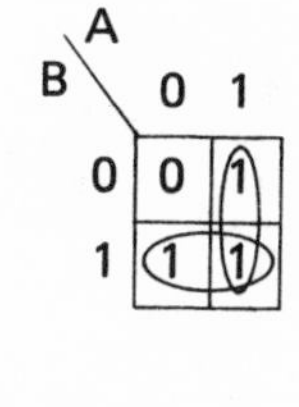

Figure 4-48

$$C = \bar{A}B + B\bar{A} + AB$$

The cube representing "AB" is common to both $A\bar{B}$ and $\bar{A}B$. (See Figure 4-48.) The switching function may be written as one term for each "circled" adjacent one, while eliminating the variable that changed.

Lower Circle

$C = B + A$ ← Circle to the right

Example:

Truth Table

A	B	C	D	Z
0	0	0	0	0
0	0	0	1	0
0	0	1	0	1
0	0	1	1	0
0	1	0	0	0
0	1	0	1	1
0	1	1	0	0
0	1	1	1	1
1	0	0	0	0
1	0	0	1	0
1	0	1	0	1
1	0	1	1	0
1	1	0	0	0
1	1	0	1	0
1	1	1	0	0
1	1	1	1	1

Karnaugh Map

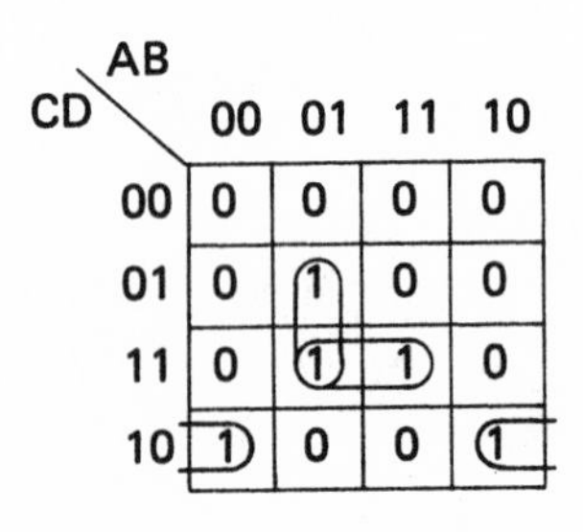

Figure 4-49

In the example in Figure 4-49, there is a common cube $\bar{A}BCD$ and a cube that is adjacent because of the "rolled" condition of Karnaugh Maps. The switching function can be written by one product term for each circle drawn on the Karnaugh Map and eliminating the variable that changed.

$$Z = \bar{A}BD + BCD + \bar{B}C\bar{D}$$

SECTION 6: EXAMPLES OF ARITHMETIC CIRCUITS

Some of the common circuits used to add and subtract circuits are given in the following examples.

Half Adder

The Half Adder is a circuit capable of giving the sum of two binary digits. (See Figure 4-50.) ·

$$\begin{array}{r} A \\ +B \\ \hline =\text{Sum} \end{array}$$

A	B	Sum
0	0	0
0	1	1
1	0	1
1	1	0

$$\text{Sum} = \bar{A}B + A\bar{B}$$

Figure 4-50

Two of the many circuits for this function are found in Figure 4-51.

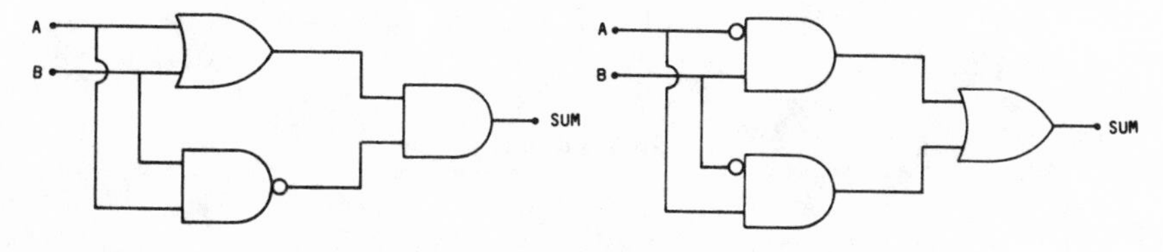

Figure 4-51

The circuit is also known as an "exclusive OR gate" and can be drawn as in Figure 4-52.

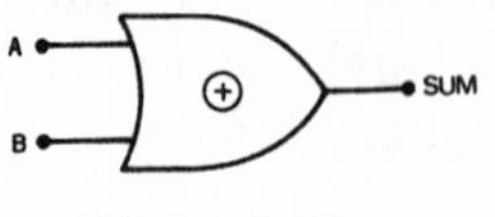

Figure 4-52

Full Adder

The Full Adder is two Half Adders plus a carry function. (See Figure 4-53.)

A	B	Sum	Carry
0	0	0	0
0	1	1	0
1	0	1	0
1	1	0	1

A
\+ B

Carry → AB

Sum → $\bar{A}B + A\bar{B}$

Figure 4-53

Half Adder with carry: (Figure 4-54.)

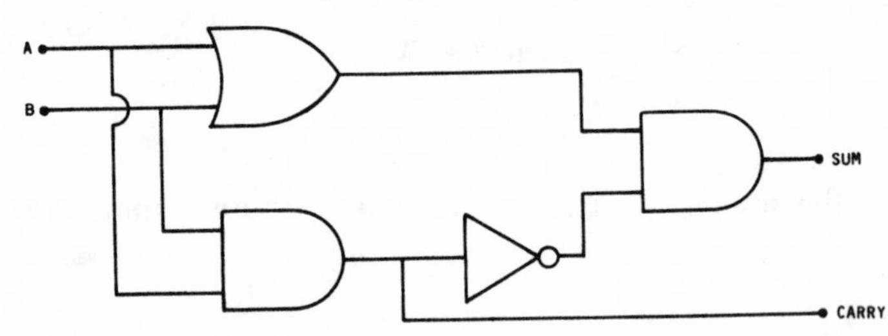

Figure 4-54

A Full Adder can be seen in Figure 4-55.

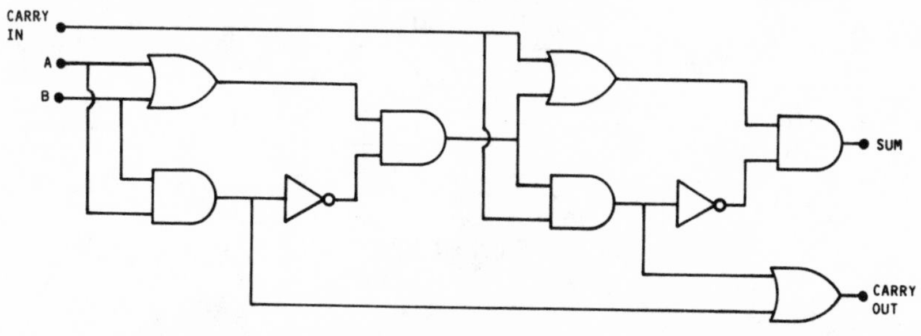

Figure 4-55

Full Adders may be cascaded to add any two sets of binary numbers. (See Figure 4-56.)

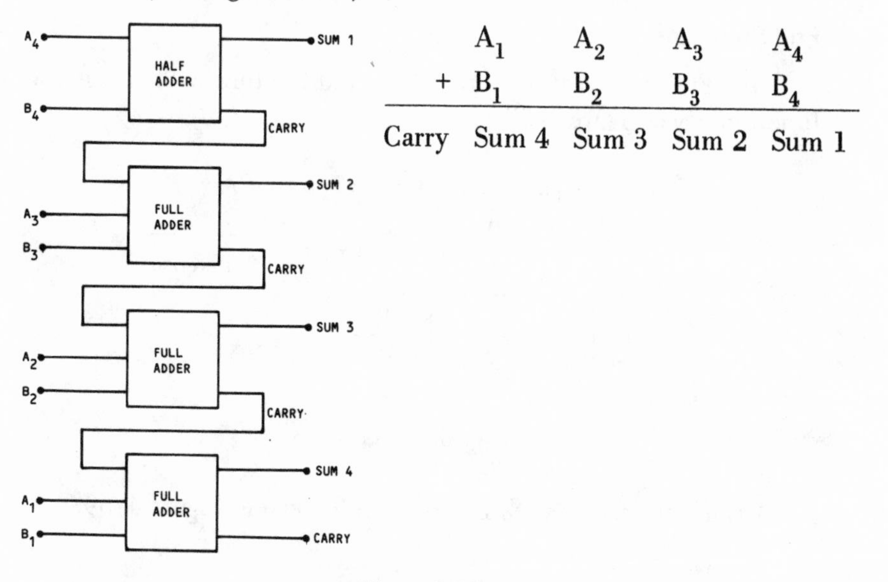

	A_1	A_2	A_3	A_4
+	B_1	B_2	B_3	B_4
Carry	Sum 4	Sum 3	Sum 2	Sum 1

Figure 4-56

Half Subtractor

The Half Subtractor is a circuit capable of giving the difference between two binary digits. (See Figure 4-57.)

A	B	Difference
0	0	0
0	1	1
1	0	1
1	1	0

Figure 4-57

Difference = $\overline{A}B + A\overline{B}$

The Half Subtractor is identical to the Half Adder.

Full Subtractor

The Full Subtractor is two Half Subtractors with a borrow function. (See Figure 4-58.)

A	B	Difference	Borrow
0	0	0	0
0	1	1	1
1	0	1	0
1	1	0	0

A
\- B
Difference

Figure 4-58

A Half Subtractor with borrow can be seen in Figure 4-59.

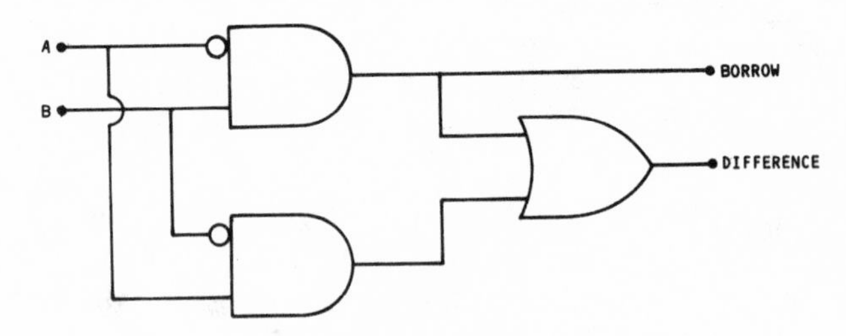

Figure 4-59

Parity

In most manipulations and transmissions of binary numbers, there exists what is called a Parity Bit. The function of the Parity Bit is to detect any errors that may have occurred due to the handling (transmission) of the numbers. There are two different types of Parity, called Even Parity and Odd Parity. Both are illustrated in the examples that follow:

Even Parity

Consider the set of binary numbers shown:

```
0 1 1 0 1
1 0 1 0 1
1 1 1 0 1
0 0 0 1 1
```

It is possible that one of the bits may change due to the noise during the transmission. To detect this possible error, a sixth column of bits is added to the right.

```
0 1 1 0 1   (1)
1 0 1 0 1   (1)
1 1 1 0 1   (0)
0 0 0 1 1   (0)
              └──→Parity Bit
```

The Parity Bit is chosen (one or zero) so that when the ones are added horizontally, the result is always an even number.

Odd Parity

With the same set of numbers, Odd Parity would be as follows:

```
0 1 1 0 1   (0)
1 0 1 0 1   (0)
1 1 1 0 1   (1)
0 0 0 1 1   (1)
              └──→Parity Bit
```

When checked in a Parity Circuit, the number shows up.

Examples of Solved Circuits

The following examples represent the basic method of deriving a logic circuit that will solve a given problem and do so with a minimum number of gates.

Circuit Design Example 1

Design a logic circuit to perform a "three-way" switch function and minimize the number of gates used.

Solution:

The first step is to generate a truth table. Let S_1 and S_2 represent switches and L represent the light or, in this case, the output. Since there are two inputs (S_1 and S_2) the combinations of inputs are shown in Figure 4-60.

A. Let 1 represent the light on and 0 represent the light off.
B. Assume the lamp is "off" when the switches are in the 0, 0 position. The lamp will then go on when one and only one switch changes position. This gives an output L as shown in Figure 4-61. It is also possible to assume the lamp is "on" with the switches in the 0, 0 position. The resulting logic circuit would be the same.

S_1	S_2	L
0	0	
0	1	
1	0	
1	1	

Figure 4-60

C. The sum-of-products switching function is:

$$L = \overline{S_1}\,S_2 + S_1\,\overline{S_2}$$

S_1	S_2	L
0	0	0
0	1	1
1	0	1
1	1	0

Figure 4-61

D. A logic circuit is seen in Figure 4-62.

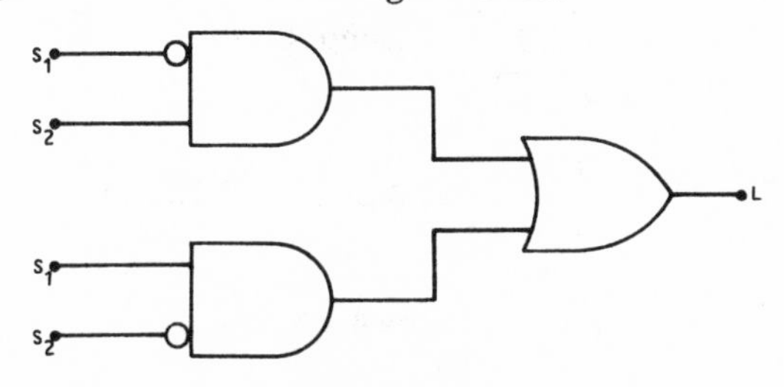

Figure 4-62

E. The two variable N-cube looks like Figure 4-63. Since there are no adjacent vertices, the circuit cannot be simplified. However, it can be rearranged.

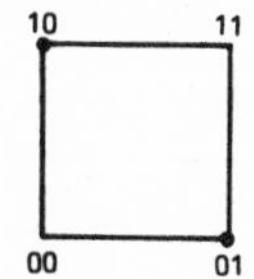

Figure 4-63

Circuit Design Example 2

Design a logic circuit to control a lamp. The lamp is to be turned on or off according to the following criteria.

1. The lamp will always be on between 6:00 p.m. and 6:00 a.m., regardless of other conditions.
2. The lamp will go off if lamp X is turned on, except per number 1 above.

Let A represent the time: *1* for 6:00 p.m. to 6:00 a.m. and *0* for daytime.
Let B represent lamp X: 1 when turned on and 0 when turned off.

A. The truth table for these conditions is as shown in Figure 4-64.

A	B	L	
0	0	1	Lamp X is off
0	1	0	
1	0	1	6:00 p.m.-6:00 a.m.
1	1	1	

Figure 4-64

B. The sum-of-product is:
$L = \overline{A}\overline{B} + A\overline{B} + AB$

C. Using Algebra to simplify:
$L = \overline{A}\overline{B} + A\overline{B} + AB$
$L = \overline{B}(\overline{A} + A) + AB$
$L = \overline{B} + AB$
$L = \overline{B} + A$ (Rule 18)

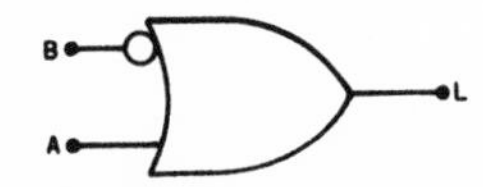

Figure 4-65

D. The result in Figure 4-65 can be derived directly from the truth table of the product of sums, which is used to derive the switching function.

5 USING TRIGONOMETRY IN ELECTRONICS

Trigonometry is the branch of math that deals with triangles. It involves the relationship between the angles within the triangle and the ratio of the lengths of the sides. These functions, derived from basic geometry, define commonly used electronic waveforms, making them very useful in electronics.

DEFINITIONS OF FUNCTIONS

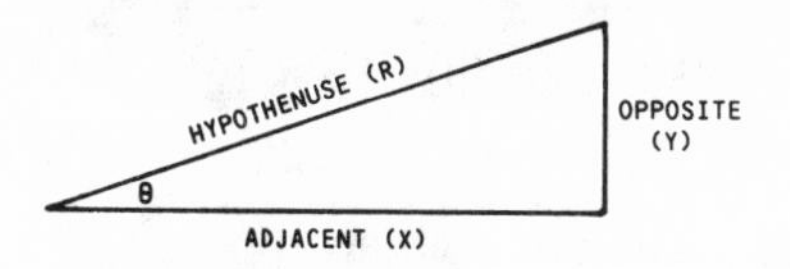

Figure 5-1

The trigonometric functions are defined as ratios of sides of a right triangle, such as in Figure 5-1. The physical size of the triangle changes the lengths of the sides but does not change the ratio of the lengths. The ratio is determined solely by the angle θ.

R ⟶ Hypotenuse: the side opposite the triangle's "right" angle.

Y ⟶ Opposite, because it is opposite the angle θ.

X ⟶ Adjacent, because it is adjacent to the angle θ.

The right triangle has the following properties:

A. One angle is 90 degrees, this is what is meant by a "right" triangle.
B. The sum of the three inside angles must add to 180 degrees.
C. The length of the sides of a right triangle are related by the Pythagorean Theorem.

$$(\text{adjacent})^2 + (\text{opposite})^2 = (\text{hypothenuse})^2 \text{ or,}$$
$$X^2 + Y^2 = R^2$$

The three basic trig functions are defined as follows:

$$\text{Sine } \theta = \frac{\text{opposite}}{\text{hypotenuse}} = \frac{Y}{R}$$

$$\text{Cosine } \theta = \frac{\text{adjacent}}{\text{hypotenuse}} = \frac{X}{R}$$

$$\text{Tangent } \theta = \frac{\text{opposite}}{\text{adjacent}} = \frac{Y}{X}$$

Three other trig functions are defined as the inverse ratios:

$$\text{Cosecant } \theta = \frac{1}{\text{sine } \theta} = \frac{\text{hypotenuse}}{\text{opposite}} = \frac{R}{Y}$$

$$\text{Secant } \theta = \frac{1}{\text{cosine } \theta} = \frac{\text{hypotenuse}}{\text{adjacent}} = \frac{R}{X}$$

$$\text{Cotangent } \theta = \frac{1}{\text{tangent } \theta} = \frac{\text{adjacent}}{\text{opposite}} = \frac{X}{Y}$$

It is convenient to imagine the right triangle inside a circle of radius "1" as shown in Figure 5-2.

The hypotenuse R may rotate anywhere within the circle, i.e., θ may have values between 0 and 360 degrees. The basic trigonometric functions now become:

$$\text{Sine } \theta = \text{Sin } \theta = \frac{Y}{R} = \frac{Y}{1} = Y$$

$$\text{Cosine } \theta = \text{Cos } \theta = \frac{X}{R} = \frac{X}{1} = X$$

$$\text{Tangent } \theta = \text{Tan } \theta = \frac{Y}{X}$$

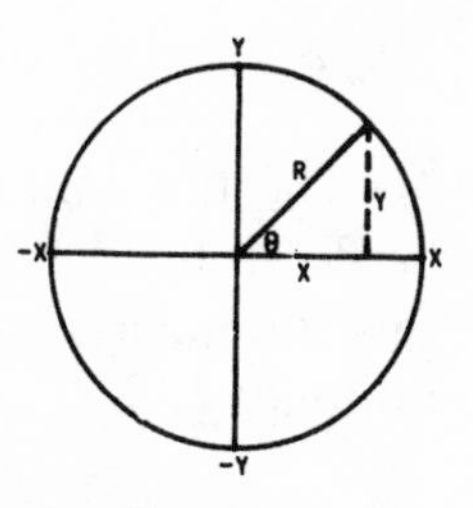

Figure 5-2

As θ varies from 0 to 360 degrees, the functions have values as shown in Figure 5-3.

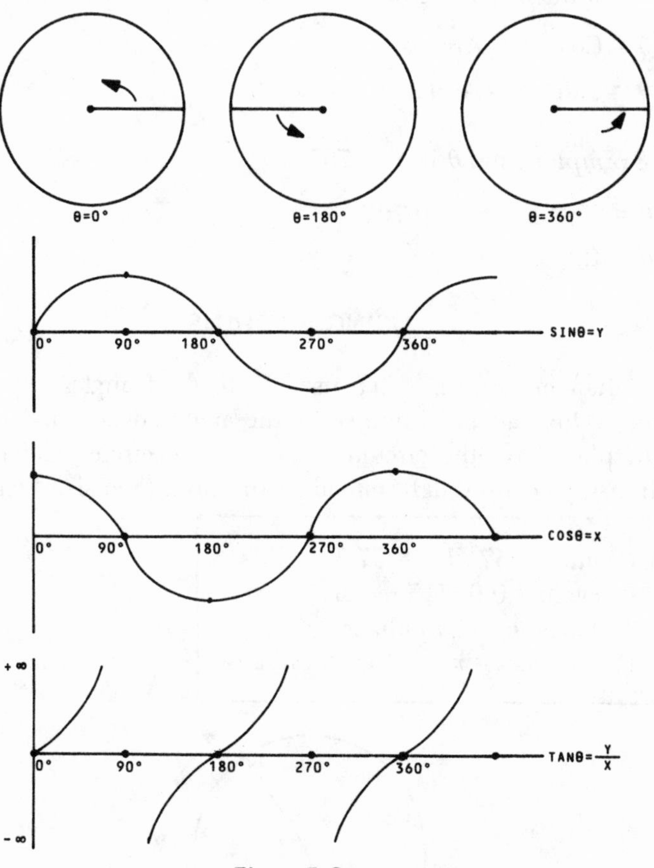

Figure 5-3

INVERSE TRIG FUNCTIONS

The function Sin θ = X says that for an angle θ, a value X can be determined that corresponds to the value sin θ. The inverse of this statement is: given a value X, determine an angle θ, which corresponds to X. For sine, this can be written:

$\theta = \text{Sin}^{-1} X$, or

$\theta = \text{Arcsin } X$

Likewise, the other trig functions have inverse functions.

$\theta = \text{Cos}^{-1} X = \text{Arccos } X$

$\theta = \text{Tan}^{-1} X = \text{Arctan } X$

Example: Find θ if X = .707

$\theta = \text{Sin}^{-1} X = \text{Sin}^{-1} 0.707$

$\theta = 45°$

USING RADIANS

Radian measure is often used instead of angles expressed in degrees. One radian is defined as the angle covered by one radial length placed on the circumference of the circle. The following relationships can be determined in the box. (See also Figure 5-4.)

1 radian = 57° 18 ≈ 57.3°
1 degree = 0.01745 radians
360 degrees = 2π radians
180 degrees = π = 3.14159 radians

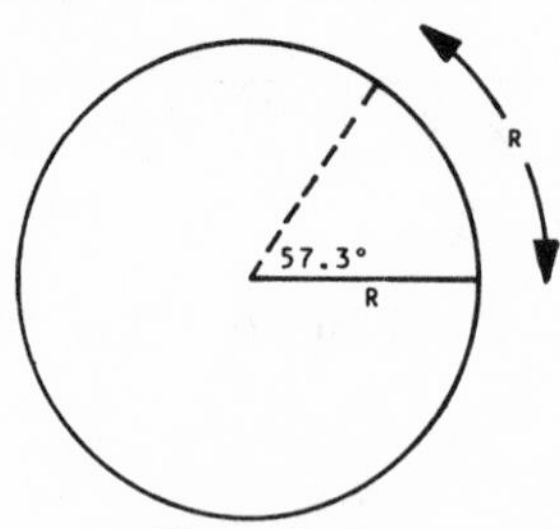

Figure 5-4

CHECKLIST OF TRIG IDENTITIES

General:

$$\text{Tan } \theta = \frac{\text{Sin } \theta}{\text{Cos } \theta} \qquad \text{Sin}^2 \theta + \text{Cos}^2 \theta = 1$$

Negative Angles:

$$\text{Cos } (-\theta) = \text{Cos } \theta \qquad \text{Sin } (-\theta) = -\text{Sin } \theta$$

Angle Addition and Subtraction:

$$\text{Sin } (A \pm B) = \text{Sin A Cos B} \pm \text{Cos A Sin B}$$
$$\text{Cos } (A \pm B) = \text{Cos A Cos B} \pm \text{Sin A Sin B}$$

90° Relationships:

$$\text{Cos } (\theta) = \text{Sin } (90 - \theta) \qquad \text{Sin } \theta = \text{Cos } (90 - \theta)$$
$$\text{Tan } (\theta) = \text{Cot } (90 - \theta) \qquad \text{Cot } \theta = \text{Tan } (90 - \theta)$$

Double Angles and Squares:

$$\text{Cos } 2\,\theta = \text{Cos}^2 \theta - \text{Sin}^2 \theta = 2\text{Cos}^2 \theta - 1$$

$$\text{Sin } 2\,\theta = 2 \text{ Sin } \theta \text{ Cos } \theta$$

$$\text{Sin}^2 \theta = \frac{1}{2}(1 - \text{Cos } 2\,\theta)$$

$$\text{Cos}^2 \theta = \frac{1}{2}(1 + \text{Cos } 2\,\theta)$$

Multiplication:

$$\text{Sin A Cos B} = \frac{1}{2}[\text{Sin } (A + B) + \text{Sin } (A - B)]$$

$$\text{Cos A Cos B} = \frac{1}{2}[\text{Cos } (A + B) + \text{Cos } (A - B)]$$

$$\text{Sin A Sin B} = \frac{1}{2}[\text{Cos (A - B) - Cos (A + B)}]$$

$$\text{Cos A Sin B} = \frac{1}{2}[\text{Cos (A + B) - Sin (A - B)}]$$

Derivatives and Integrals:

$$\frac{d \text{ Sin } \theta}{d\theta} = \text{Cos } \theta \qquad \frac{d \text{ Cos } \theta}{d\theta} = -\text{Sin } \theta$$

$$\int \text{Sin } \theta \, d\theta = -\text{Cos } \theta + k \qquad \int \text{Cos } \theta \, d\theta = \text{Sin } \theta + k$$

Most Commonly Used Values:

DEGREES	0	30	45	60	90	180	270	360
RADIANS	0 0.0	$\pi/6$ 0.524	$\pi/4$ 0.785	$\pi/3$ 1.047	$\pi/2$ 1.571	π 3.142	$3\pi/2$ 4.712	2π 6.283
SIN	0	$1/2$	$\sqrt{2}/2$	$\sqrt{3}/2$	1	0	-1	0
COS	1	$\sqrt{3}/2$	$\sqrt{2}/2$	$1/2$	0	-1	0	1

Table 5-1

HOW TO APPLY TRIG TO ALTERNATING CURRENT

The most common waveform in electronics is the sine wave. It is so called because its voltage takes the shape of the trigonometric function sine θ. The general expression for an A.C. voltage is found in Figure 5-5.

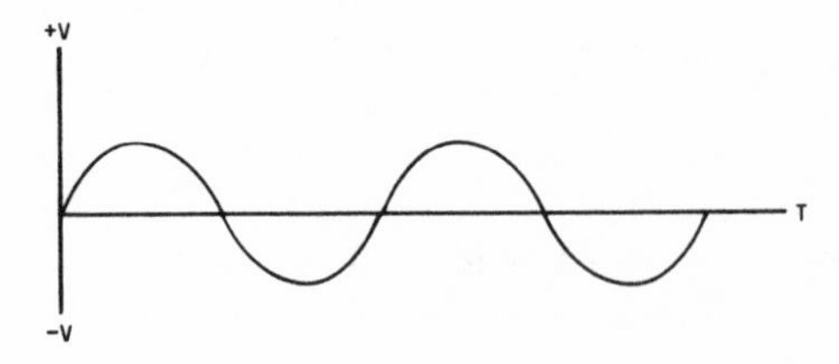

Figure 5-5

$V(t) = A \sin (2\pi ft + \theta)$

Where: (A) is the peak voltage.
(f) is the frequency in cycles per second (Hertz).
(θ) is the phase angle of the voltage compared to some reference.
(t) is time in seconds.

The equation may also be written:

$V(t) = A \sin (wt + \theta)$

Where: $w = 2\pi f$ is the angular frequency in radians per second

Peak Average and RMS:

Since the voltage of a sine wave is constantly changing, it is necessary to define what is meant by A.C. voltage and also to explain how to use A.C. voltage in Ohm's Law and other formulae. Figure 5-6 shows one cycle of an A.C. waveform and the definition of the voltages associated with it.

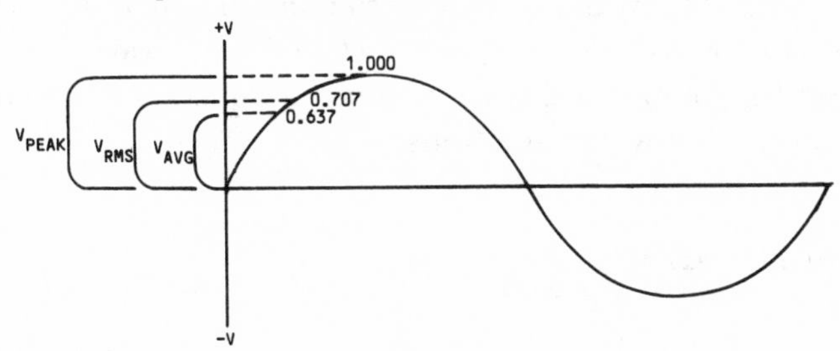

Figure 5-6

Vpeak—The peak voltage is the value from zero to the maximum positive or negative voltage of the waveform.

Vp-p— The peak-to-peak voltage is twice the peak voltage $Vp\text{-}p = 2Vp$

Vavg— The average voltage is the average of one-half the sinewave. It is equal to $0.637 \times Vp$.

Vrms— The R.M.S. (root mean square) voltage is the value used in Ohms Law and other formula. It is equal to $0.707 \times Vp$.

Frequency, Period, and Phase Angle:

A sinewave goes through one cycle when starting at zero; it goes positive, negative, and back to zero as shown in Figure 5-7. One cycle of a sinewave can also be determined from two points, such as point A to point B, and point C to point D. One cycle is also:

A. 1 cycle = 360 degrees
B. 1 cycle = 2π radians

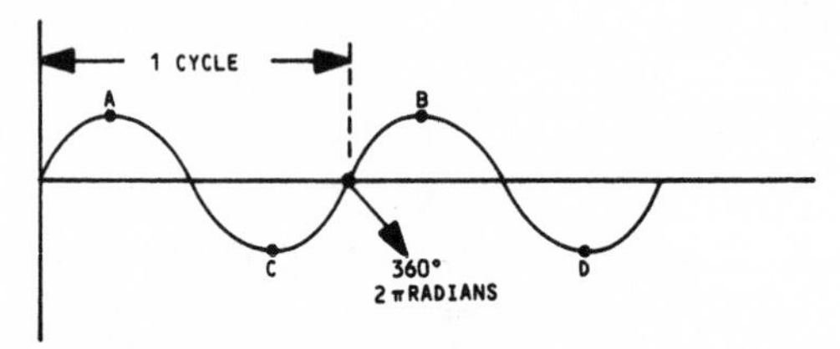

Figure 5-7

The *period(T)* of a sinewave is the amount of time it takes to complete one cycle. The *frequency (F)* of a sinewave is the number of cycles completed in one second. Frequency and period are related by the following equations:

$$\text{frequency (HZ)} = \frac{1}{\text{period}}$$

$$\text{period} = \frac{1}{\text{frequency (HZ)}}$$

Example: If the period of a sinewave is 0.01 seconds, what is the frequency?

$$\text{f(HZ)} = \frac{1}{\text{period}} = \frac{1}{.01} = \textit{100 Hertz}$$

The sinewave has 100 cycles every second or a frequency of 100 HZ.

Example: What is the period of a 9 megahertz sinewave?

$$T = \frac{1}{F} = \frac{1}{9 \times 10^{+6}} = 0.1111 \times 10^{-6} \text{ sec.}$$

$$T = 0.1111 \text{ microseconds}$$

Phase is the term used to describe the angular relationship between two or more sinewaves. Consider the two waveforms in Figure 5-8. A sinewave and a cosine wave are 90 degrees apart, or to put it another way, the phase relationship between the two waveforms is 90 degrees.

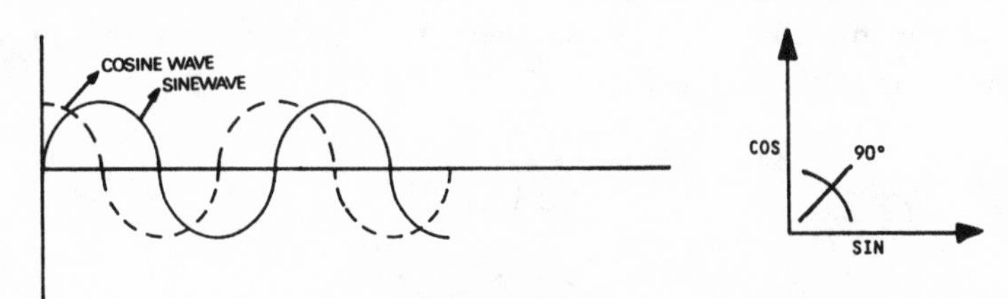

Figure 5-8

Using the j Operator:

All real numbers can be represented by a line as shown in Figure 5-9.

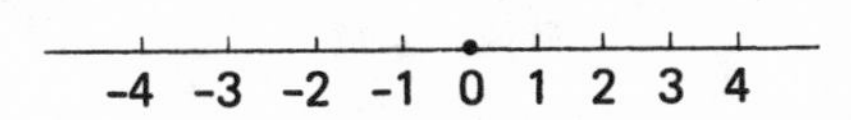

Figure 5-9

Negative numbers are on the left side of zero and positive numbers are on the right side. Rational numbers such as 3, and irrational numbers such as $\sqrt{2}$ can all be represented by a point on this line. Numbers that cannot be represented by this line are called *imaginary* numbers. A good example of this is the number $\sqrt{-1}$. Since (positive) $\times$ (positive) = positive, and (negative) $\times$ (negative) = positive, it is very difficult to conceive of a number that, when squared, equals (-1).

To understand imaginary numbers such as $\sqrt{-1}$, it is convenient to think of the number line as in Figure 5-10.

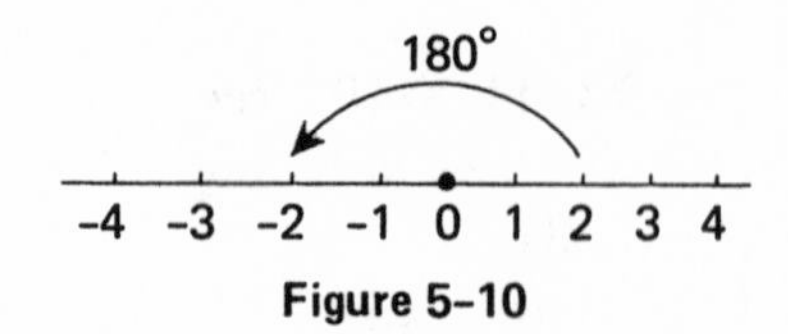

Figure 5-10

If zero is taken as the pivot point, negative numbers can be considered as having been rotated 180° with respect to positive numbers; or, said another way, a negative sign (-) applied to a number implies 180° rotation of that number.

The j operator, when applied to a number, implies a 90° rotation of the number. The j axis can then be shown with the real number axis as shown in Figure 5-11.

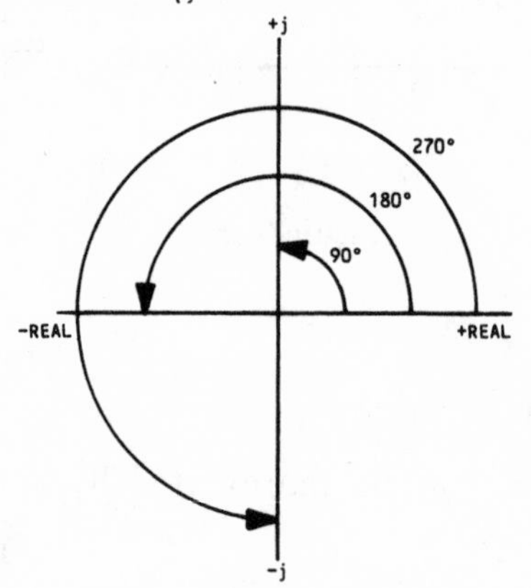

Figure 5-11

By definition, j is equal to the square root of -1, ($\sqrt{-1}$). When j is applied to a number, it says the number has been rotated +90 degrees. Since j equals $\sqrt{-1}$, j^2 equals -1, so the -1 operator implies rotation by 180 degrees, or rotation to negative numbers. Thus, j^3 equals $-\sqrt{-1}$ and implies a shift of 270 degrees, or a shift of -90 degrees. These characteristics of j are summarized as follows:

$0° = +1$

$90° = j = \sqrt{-1}$

$180° = j^2 = -1$

$270° = j^3 = -\sqrt{-1}$

$360° = j^4 = +1$ (Same as 0°)

COMPLEX NUMBERS

Complex numbers are formed by combinations of real and imaginary numbers. Usually the real number is written first. As an example, $(5 + 9j)$ is a complex number representing 5 units on the positive real axis, plus 9 units on the positive j axis, as shown in

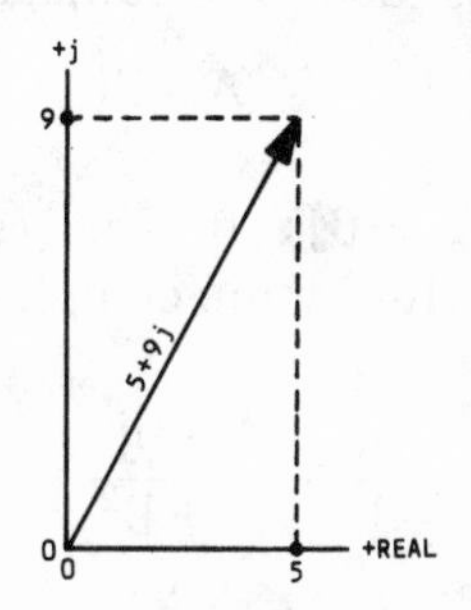

Figure 5-12

Figure 5-12. Arithmetic operations can be performed using complex numbers as well as real numbers. For example:

Addition

$$\begin{array}{c} A + Bj \\ M + Cj \\ \hline = (A + M) + (B + C)j \end{array} \qquad \begin{array}{c} A + Bj \\ -D - Cj \\ \hline = (A - D) + (B - C)j \end{array}$$

Subtraction

$$\begin{array}{c} A + Bj \\ \underline{-(D + Ej)} \end{array} \; = \; \begin{array}{c} A + Bj \\ -D - Ej \\ \hline = (A - D) + (B - E)j \end{array}$$

Multiplication

$$\begin{array}{c} A + Bj \\ \times\, M + Cj \\ \hline \end{array}$$

$$AM + BMj + ACj + BCj^2$$

$$= AM + (BM + AC)j + BCj^2 \longrightarrow j^2 = -1$$

$$= (AM - BC) + (BM + AC)j$$

Division

To divide complex numbers, it is necessary to use the *complex conjugate.* The complex conjugate is formed by taking the complex number and changing the sign of the j quantity.

	Number	*Complex Conjugate*
Example:	A + Bj	A - Bj

To perform 1 ÷ (A + BJ), it is necessary to multiply both numerator and denominator by the complex conjugate.

$$\frac{1}{A + Bj} = \frac{1}{A + Bj} \times \frac{A - Bj}{A - Bj} = \frac{A - Bj}{A^2 + B^2}$$

$$\frac{1}{A + Bj} = \left(\frac{A}{A^2 + B^2}\right) - \left(\frac{B}{A^2 + B^2}\right)j$$

In numbers instead of letters:

$$\frac{1}{2 + 3j} = \frac{1}{2 + 3j} \times \frac{2 - 3j}{2 - 3j} = \frac{2 - 3j}{4 + 9}$$

$$= \frac{2}{13} - \frac{3}{13}j$$

APPLICATION TO CIRCUITS

In an electronic circuit, A (+j) is applied to inductive reactance, while (-j) is applied to capacitive reactance. Circuit impedance can then be represented as in Figure 5-13.

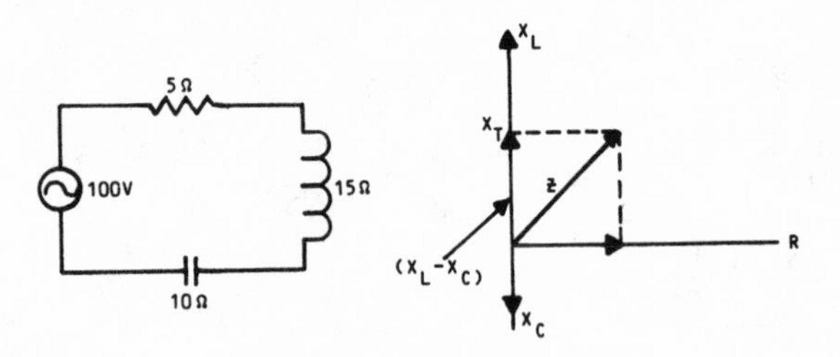

Figure 5-13

$$Z = 5 + 15j - 10j$$
$$= 5 + 5j$$

$$I = \frac{E}{Z} = \frac{100}{5 + 5j} = \frac{20}{1 + 1j} = \frac{20}{1 + 1j} \times \frac{1 - 1j}{1 - 1j}$$

$$= \frac{20 - 20j}{2} = 10 - 10j$$

Examples of Solved Circuits

The following contains several examples of series, parallel and series/parallel circuits where an analysis of the circuit has been done using trigonometry and the j operator. There are special examples on converting a parallel circuit to an equivalent series circuit. (See Figure 5-14.)

Example 1: Series Circuit

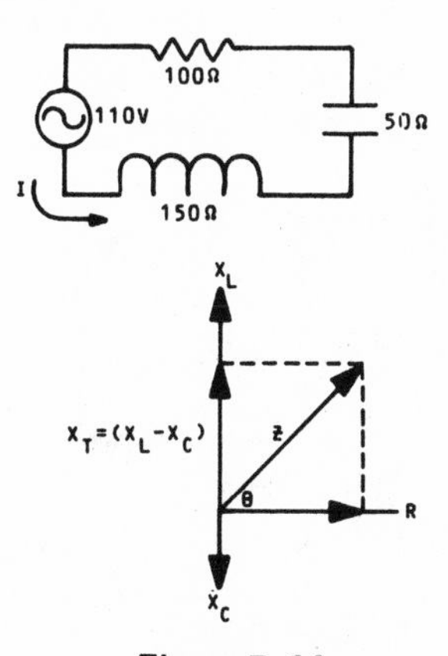

Figure 5-14

By Trig:

$$X_T = 100\Omega X_L$$
$$R = 100\Omega$$

$$\text{Tan } \theta = \frac{X}{R} = \frac{100}{100} = 1$$

$\theta = 45°$ E leads

$$\text{Sin } \theta = \frac{X}{Z} \quad \text{or} \quad Z = \frac{X}{\text{Sin } \theta}$$

$$Z = \frac{100}{0.707} = 141.4 \text{ ohms}$$

$$I = \frac{E}{Z} = \frac{100}{141.4} = 0.707 \text{ amps}$$

$E_R = IR = 0.707 \times 100 = 70.7$ volts

$E_C = IX_C = 0.707 \times 50 = 35.35$ volts

$E_L = IX_L = 0.707 \times 150 = 106.05$ volts

E (source) must equal $E_R + E_C + E_L$ (Kirchoff's Law)

The three are added as vectors in Figure 5-15.

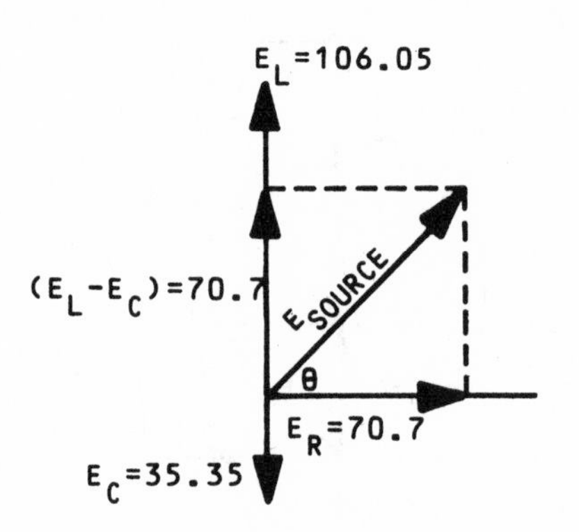

Figure 5-15

$$\theta = \text{Tan}^{-1}\left(\frac{70.7}{70.7}\right)$$

$\theta = \text{Tan}^{-1}$ (1)

$\theta = 45°$

$$\text{E source} = \frac{70.7}{\sin \theta} = \frac{70.7}{.707} = 100 \text{ volts}$$

By the j Operator

$$Z = 100 + 100j$$

$$I = \frac{E}{Z} = \frac{100}{100 + 100j}$$

$$I = \frac{100}{100 + 100j} \times \frac{100 - 100j}{100 - 100j} \quad \text{(multiply by complex conjugate)}$$

$$I = \frac{10000 - 10000j}{10000 - 10000j^2} = \frac{10000 - 10000j}{10000 + 10000} = \frac{10000 - 10000j}{20000}$$

$$I = 0.5 - 0.5j \quad \text{(rectangular form)}$$

$$I = 0.707 \angle{-45} \quad \text{(polar form)}$$

$$E_R = IR = (0.5 - 0.5j) \times 100 = +50 - 50j$$

$$E_C = IX_C = (0.5 - 0.5j) \times -50j = -25 - 25j$$

$$E_L = IX_L = (0.5 - 0.5j) \times 150j = +75 = 75j$$

$$E_S = E_R + E_C + E_1 \qquad \begin{array}{r} +50 - 50j \\ -25 - 25j \\ +75 + 75j \\ \hline \end{array}$$

$$E_S = 100 \text{ volts}$$

Example 2: Series Circuit (Figure 5-16)

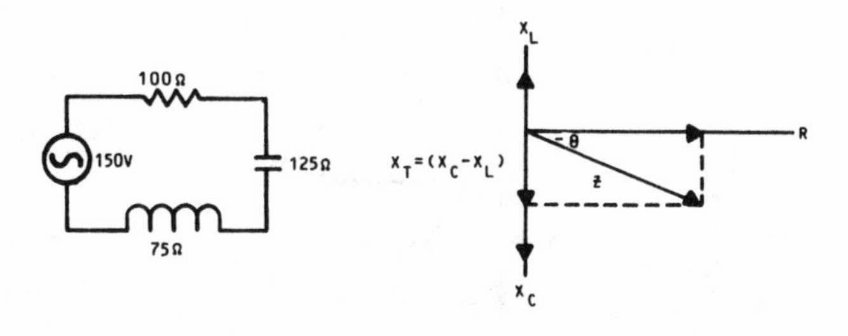

Figure 5-16

By Trig

$X_T = 50\Omega X_C$

$R = 100\Omega$

$$\text{Tan}\ \theta = \frac{-X}{R} = \frac{-50}{100} = -0.5$$

$\theta = \text{Tan}^{-1}\ (-0.5) = -26.56°$ *I leads*

$$Z = \frac{X}{\text{Sin}\ \theta} = \frac{-50}{-0.4471} = 111.823 \text{ ohms}$$

$$I = \frac{E}{Z} = \frac{150}{111.823} = 1.341 \text{ amps}$$

$E_R = IR = 1.341 \times 100 = 134.1$ volts

$E_C = IX_C = 1.341 \times 125 = 167.625$ volts

$E_L = IX_L = 1.341 \times 75 = 100.575$ volts

E source $= E_L + E_C + E_R$

The three are added as vector in Figure 5-17:

$$\theta = \text{Tan}^{-1} \frac{67.05}{134.1} - 26.56°$$

$$\text{E source} = \frac{67}{\text{Sin}\ \theta} = 150 \text{ volts}$$

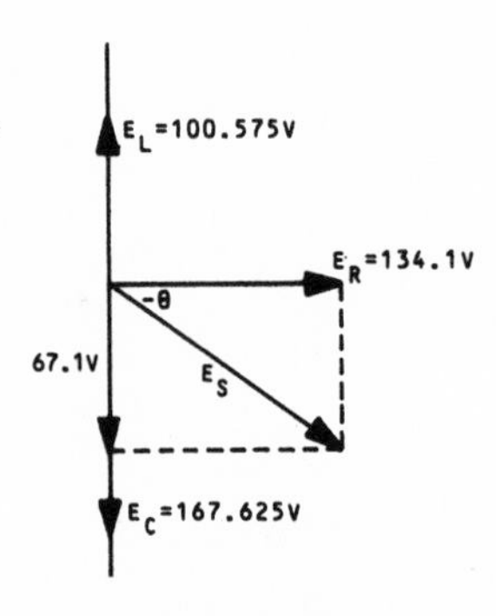

Figure 5-17

By the j Operator

$$Z = 100 - 50j$$

$$I = \frac{E}{Z} = \frac{150}{100 - 50j}$$

$$I = \frac{150}{100 - 50j} = \frac{1.5}{1 - 0.5j} \quad \text{(divide by 100)}$$

$$I = \frac{1.5}{1 - 0.5j} \times \frac{1 + 0.5j}{1 + 0.5j} \quad \text{(multiply by complex conjugate)}$$

$$I = \frac{1.5 + 0.75j}{1 + 0.25} = \frac{1.5 + 0.75j}{1.25}$$

$$I = 1.2 + 0.6j \quad \text{(rectangular)}$$

$$\theta = \tan^{-1}\frac{0.6}{1.2} = 26.56°$$

$$I_{magnitude} = \frac{0.6}{\sin\theta} = 1.342 \text{ amps}$$

$$I = 1.342 \angle 26.56 \quad \text{(polar)}$$

$$E_R = IR = (1.2 + 0.6j) \times 100 = +120 + 60j$$

$$E_C = IX_C = (1.2 + 0.6j) \times -125j = +75 - 150j$$

$$E_L = IX_L = (1.2 + 0.6j) \times +75j = \underline{-45 + 90j}$$

$$E_S \qquad = 150$$

Example 3: Parallel Circuit (Figure 5-18.)

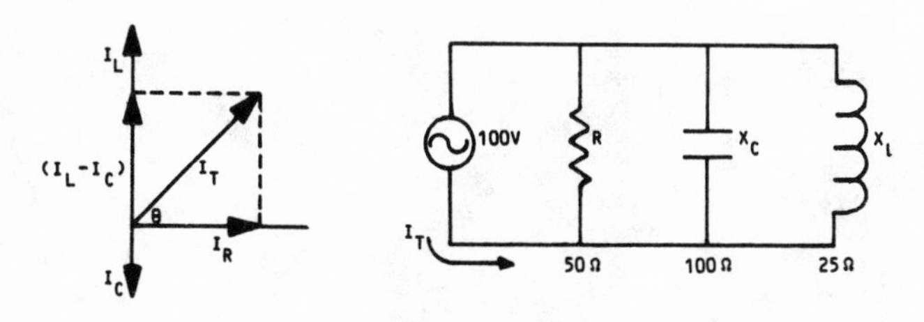

Figure 5-18

By Trig

$$I_R = \frac{E}{R} = \frac{100}{50} = 2\text{A}$$

$$I_C = \frac{E}{X_C} = \frac{100}{100} = 1\text{A}$$

$$I_L = \frac{E}{X_L} = \frac{100}{25} = 4\text{A}$$

$I_T = I_R + I_C + I_L$ (by Kirchoff's Law)

I_T is shown in Figure 5-18. It is the vector sum of I_R, I_C and I_L.

$$\theta = \text{Tan}^{-1} \frac{I_L - I_C}{I_R} = \frac{3}{2} = 56.31°$$

$$\text{Sin } \theta = \frac{I_L - I_C}{I_T} \qquad I_T = \frac{I_L = I_C}{\text{Sin } \theta} = \frac{3}{\text{Sin } \theta}$$

$I_T = 3.61$ amps

$$Z = \frac{E}{I} = \frac{100}{3.61} = 27.7 \text{ ohms}$$

By the j Operator

In this example, we will convert resistance, reactance, and impedance into conductance, susceptance, and admittance to solve for the circuit parameters.

$$\text{Conductance} = G = \frac{1}{R}$$

$$\text{Susceptance} = B = \frac{1}{\pm X}$$

$$\text{Admittance} = Y = \frac{1}{Z}$$

$$G = \frac{1}{R} = \frac{1}{50} = 0.02$$

$$B_L = \frac{1}{X_L} = \frac{1}{25j} = \frac{1}{25j} \times \frac{-25j}{-25j} = \frac{-25j}{625} = -0.04j$$

$$B_C = \frac{1}{X_C} = \frac{1}{-100j} = \frac{1}{-100j} \times \frac{100j}{100j} = \frac{100j}{10000} = 0.01j$$

$$\theta = \text{Tan}^{-1} \frac{B_L - B_C}{G} = 56.31°$$

$$Y = \frac{B_L - B_C}{\text{Sin } \theta} = 0.036 \angle 56.31° \qquad \text{(polar)}$$

$$Y = 0.02 - 0.03j \qquad \text{(rectangular)}$$

$$I = \frac{E}{Z} = EY$$

$$I = 100(0.02 - 0.03j)$$

$$I = 2 - 3j \qquad \text{(rectangular)}$$

$$\theta = \text{Tan}^{-1} \frac{-3}{2} = -56.31°$$

$$I = \frac{-3}{\text{Sin } (-56.31)} = 3.61 \text{ amps}$$

$$Z = \frac{E}{I} = \frac{100}{2 - 3j}$$

$$Z = \frac{100}{2 - 3j} \times \frac{2 + 3j}{2 + 3j} = \frac{200 + 300j}{4 + 9} = 15.38 + 23.08j$$

$$Z = 15.38 + 23.08J \qquad \text{(rectangular)}$$

$$Z = \frac{23.08}{\text{Sin } \theta} = 27.7 \text{ ohms}$$

Example 4: Converting Parallel to Series

Given the parallel circuit in Figure 5-19, find a series circuit that gives the same impedance.

$$Z = \frac{Z_1 Z_2}{Z_1 + Z_2} = \frac{8 \times 12j}{8 + 12j}$$

$$Z = \frac{+96j}{8 + 12j} = \frac{+24j}{2 + 3j}$$

Figure 5-19

$$\frac{+24j}{2 + 3j} \times \frac{2 - 3j}{2 - 3j} = \frac{+48j + 72}{4 + 9} = \frac{72 + 48j}{13}$$

$$Z = 5.538 + 3.692j$$

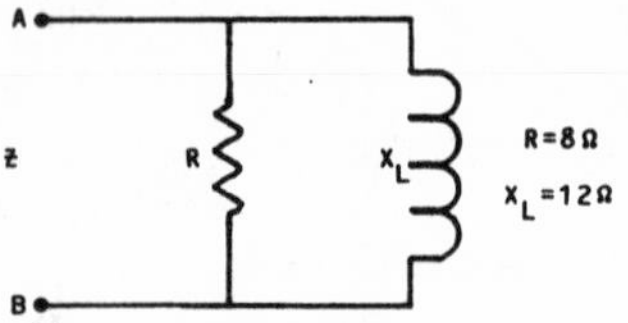

This gives a series circuit found in Figure 5-20.

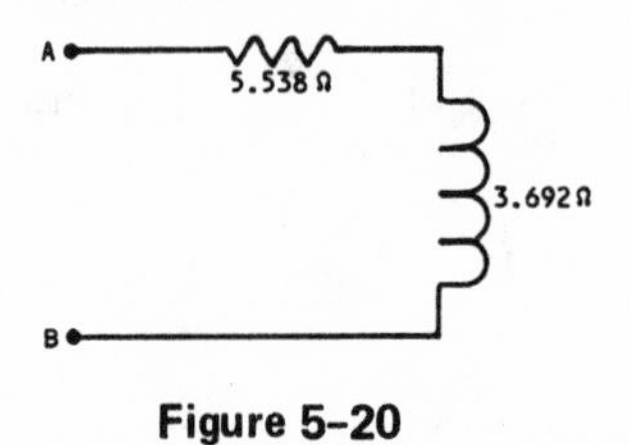

Figure 5-20

Example 5: Series Parallel Circuit

Find the impedance between A and B, in Figure 5-21.

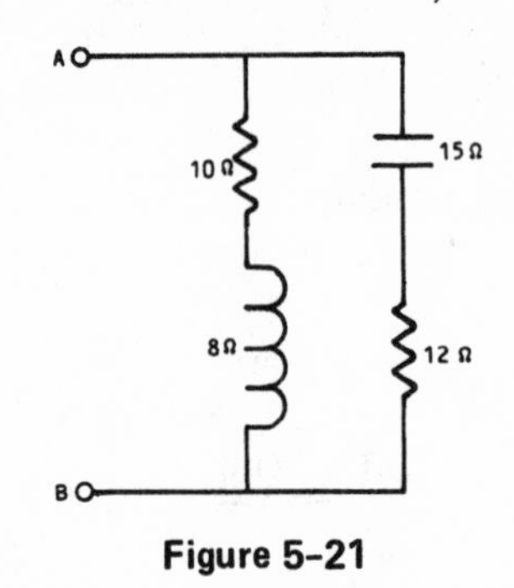

Figure 5-21

$$Z_1 = 10 + 8j \qquad Z_2 = 12 - 15j$$

$$Z_T = \frac{Z_1 Z_2}{Z_1 + Z_2}$$

$$Z = \frac{(10 + 8j)(12 - 15j)}{(10 + 8j) + (12 - 15j)} = \frac{240 - 54j}{22 - 7j}$$

$$\frac{240 - 54j}{22 - 7j} \times \frac{22 + 7j}{22 + 7j} = \frac{5658 + 492j}{484 + 49}$$

$$Z_T = 10.62 + .923j \quad \text{(rectangular)}$$

The series equivalent is found in Figure 5-22.

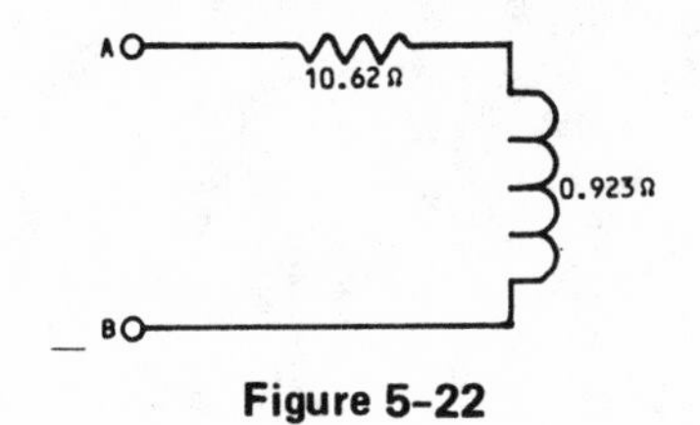

Figure 5-22

To convert to polar:

$$Z_T = 10.62 + .923j$$

$$\theta = \text{Tan}^{-1} \frac{X}{R} = \text{Tan}^{-1} \frac{.923}{10.62} = 4.967°$$

$$Z \text{ magnitude} = \frac{X}{\text{Sin } \theta} = \frac{.923}{\text{Sin } (4.967)} = 10.66$$

$$Z \text{ polar} = 10.66 \angle 4.967°$$

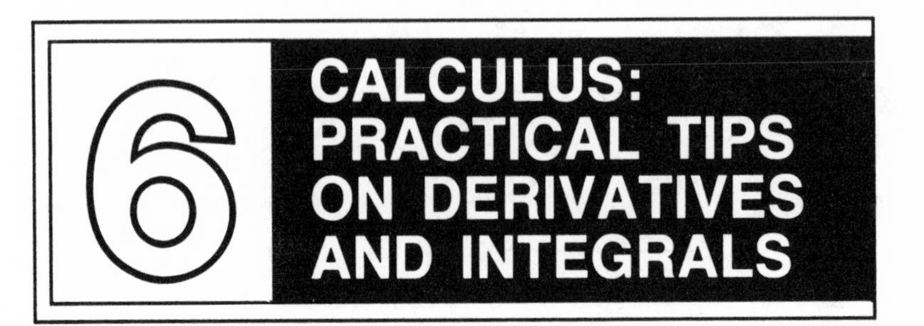

INTRODUCTION

Differential and integral calculus are very powerful tools in the solution of math problems, especially in electronics. This chapter defines the basic operations of calculus and then gives several examples of their application to electronic problems.

SECTION 1: DEFINITION OF DERIVATIVES AND INTEGRALS

Derivatives. The purpose of a derivative is to determine the instantaneous rate of change between two variables. If these two variables are related by an equality and this function is continuous, the instantaneous rate of change can be determined by calculating the first derivative. The derivative is then the slope of the function at any chosen point.

For example, assume we want to determine the slope of the $Y = F(X)$ at the point $X = X_1$, as shown in Figure 6-1. The slope of the curve is the same as the slope of the tangent line at point X_1. The slope of this tangent can be found graphically, as shown in part B of Figure 6-1.

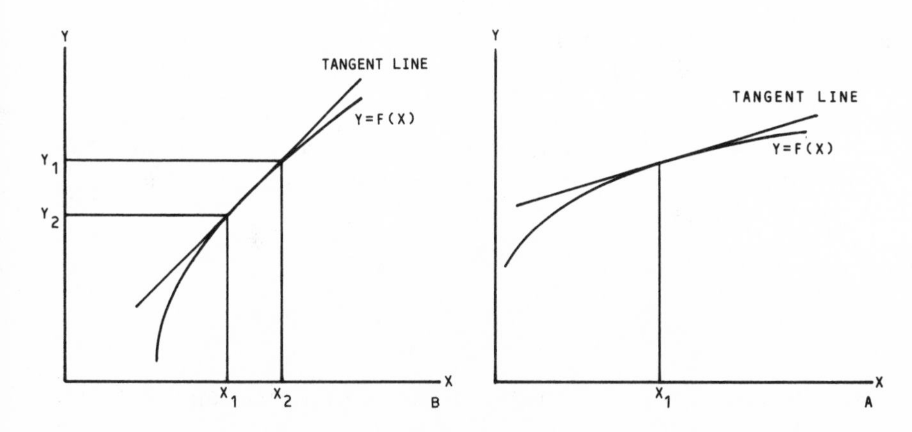

Figure 6-1

$$\text{Slope} = \frac{Y_1 - Y_2}{X_2 - X_1}$$

The purpose of a derivative is to find this slope without drawing a graph, and using the method shown in Figure 6-1.

To show how a derivative is found, consider Figure 6-2. The slope of the tangent line can be approximated by lines joining points on the function Y = F(X). The line joining F(X) at X_1 and X_2 is a first approximation of the tangent line slope. This first approximation is:

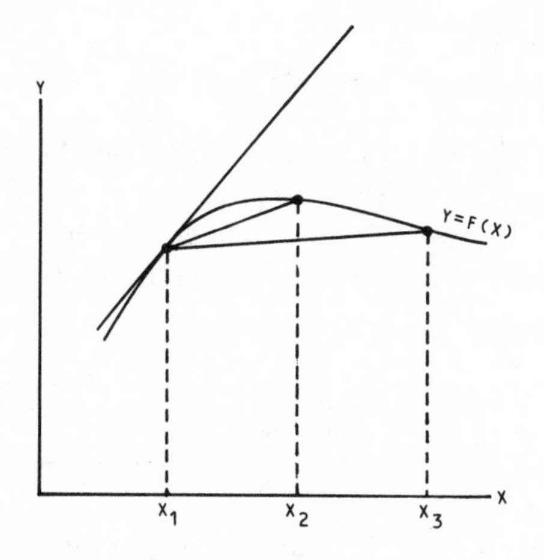

Figure 6-2

$$\text{slope} = \frac{F(X_2) - F(X_1)}{X_2 - X_1}$$

A closer approximation can be found by choosing a point (X_3) that is closer to X_1. The slope of the line between X_1 and X_3 is much closer to the slope of the tangent line.

$$\text{slope} = \frac{F(X_3) - F(X_1)}{X_3 - X_1}$$

It is this process by which the derivative of a function is defined. If the above reasoning is followed, the actual slope of the tangent line can be found if we pick a point (X_n) that is extremely close to (X_1), or a point (X_n) such that (X_n - X_1) is extremely small.

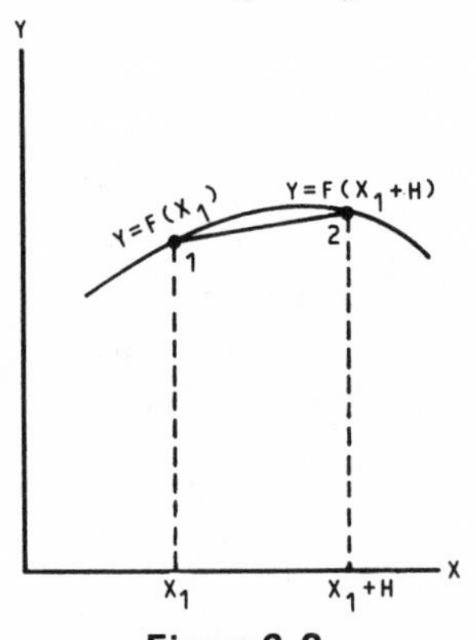

Figure 6-3

In Figure 6-3, two points are shown on the curve. The line connecting these points very closely approximates the slope of the curve at point 1. As "h" gets smaller, the approximation gets better. The derivative is defined as shown in the table.

$$\frac{dY}{dX} = \text{derivative} = \lim_{h \to 0} \frac{f(x + h) - f(x)}{h}$$

To illustrate the use of this definition, consider the following example:

Example 6–1: The energy stored in a capacitor is given by the equation: $W = \frac{CE^2}{2}$

Where: W = energy in joules
C = capacitance in farads
E = applied voltage

Find the rate of change between energy and voltage at the point E = 2.0 volts. The rate of change is the derivative $\frac{dW}{dE}$

$$\frac{dW}{dE} = \frac{f(x+h) - f(x)}{h} \quad \text{as } h \longrightarrow \text{zero}$$

$$= \frac{\frac{C(E+h)^2}{2} - \frac{CE^2}{2}}{h} = \frac{\frac{C(E^2 + 2Eh + h^2)}{2} - \frac{CE^2}{2}}{h}$$

$$= \frac{2CEh + h^2}{2h} = \left(CE + \frac{h}{2}\right)$$

As h goes to zero $(CE + \frac{h}{2})$ goes to CE, so:

$$\frac{dW}{dE} = CE \quad \text{at } E = 2.0$$

$$\frac{dW}{dE} = 2C$$

These results are shown graphically in Figure 6–4.

The slope of the curve $W = \frac{CE^2}{2}$ at the point E = 2.0 is 2.0C

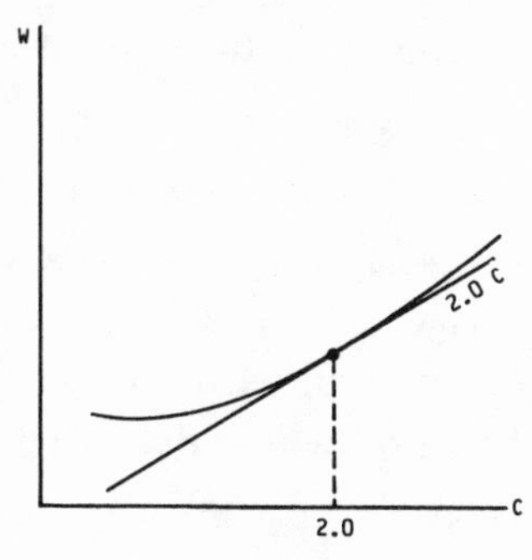

Figure 6–4

Finding derivatives by use of the definition is cumbersome and sometimes quite difficult. It is seldom if ever used. The derivative in the above example can be found from the following basic relationship.

If $Y = X^n$

$$\frac{dY}{dX} = NX^{N-1}$$

applied to: $W = \frac{CE^2}{2}$

$$\frac{dW}{dE} = \frac{2\ CE^1}{2} = CE$$

Example 6–2: The relationship between charge and current is:

$$I = \frac{dQ}{dt} \quad \text{where:}$$

I = current in amperes

Q = charge in coulombs

t = time in seconds

If $Q = \frac{T^3}{3}$, find I

$$I = \frac{dQ}{dt} = \frac{3t^2}{3} = t^2$$

If the charge is changing by $\frac{t^3}{3}$, the current is changing by t^2.

Intergrals

An integral is the opposite of a derivative. Its sign is $\int$, and $\int xdx$ is read, "the integral of x,dx". To evaluate the expression $\int xdx$, it is necessary to find a function f(x) such that $\frac{df(x)}{dx} = x$. The solution of $\int xdx$ is a function whose derivative is equal to x.

Example 6–3: $f(x) = x^2$

$$\frac{df(x)}{dx} = 2x$$

Therefore:

$[\int 2xdx]$ must be x^2

Integration is the Reverse of Differentiation.

Just as the derivative can be interpreted graphically as the slope of the curve at any given point, the integral can also be shown graphically. Consider the graph of the function $Y = X^2$ in Figure 6-5.

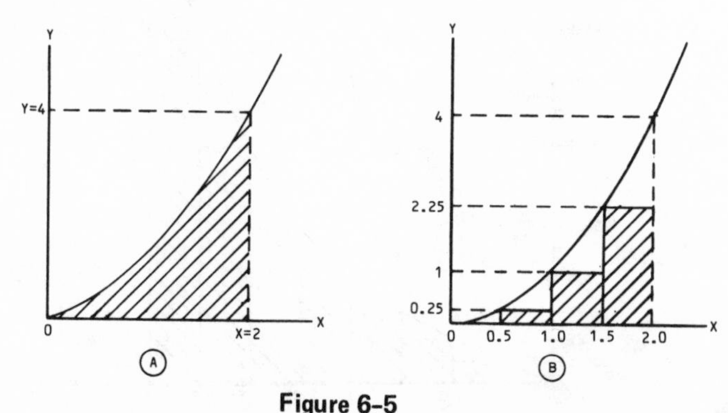

Figure 6-5

The area under the curve between X = 0 and X = 2 is shown in the shaded portion of Figure 6-5A. If we want to calculate this area, we can start as a first approximation as shown in Figure 6-5B. Divide the area into smaller portions and add their areas. In Figure 6-5B, the three shaded areas approximate the total area and this can be calculated.

$$\begin{aligned}\text{First approximation} &= (0.5 \times 0.25) + (0.5 \times 1) + (0.5 \times 2.25)\\ &= 0.125 + 0.5 + 1.125\\ &= 1.75 \text{ square units}\end{aligned}$$

A closer approximation is possible if the number of divisions of the X axis is increased. Consider Figure 6-6. Here the X axis is divided into seven columns. The total area of the rectangular columns is:

$$\begin{aligned}\text{area} = {} & (.25 \times 0.0625) + (.25 \times 0.25) + (.25 \times 0.5625)\\ & + (.25 \times 1.0) + (.25 \times 1.5625) + (.25 \times 2.25)\\ & + (.25 \times 3.0625)\\ & 0.015625 + 0.0625 + 0.140625 + 0.25\end{aligned}$$

$+ 0.390625 + 0.5625 + 0.765625$

Second approximation = 2.1875 square units

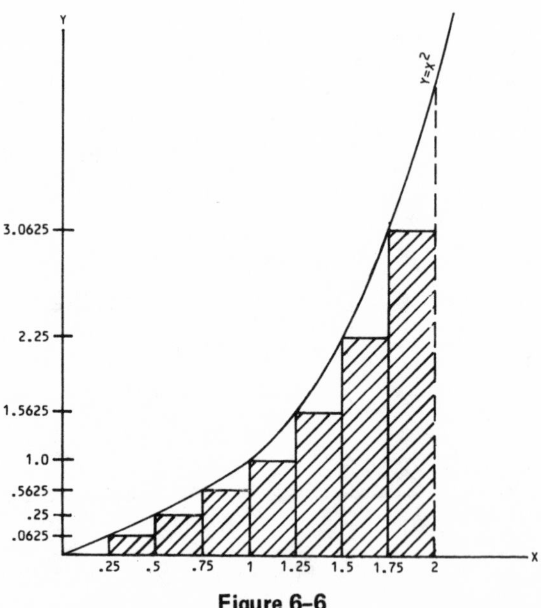

Figure 6-6

The second approximation is closer to the actual area under the curve than the first approximation because the number of divisions on the X-axis was larger. If the number of divisions on the X-axis approaches infinity, the approximation of the area under the curve approaches the real value.

$$\text{Approximate area} = \Sigma\,(A_1 + A_2 + \ldots . A_n)$$

$$\text{Actual area} = \lim_{N \to \infty} \Sigma\,(A_1 + A_2 + \ldots A_n)$$

The integral of a function, evaluated between the limits, is also the area under the curve.

Definition

$$\int_A^B f(x)dx = \lim_{N \to \infty} \Sigma\,(A_1 + A_2 + A_3 + \ldots .. A_n)$$

To apply this to the area under the curve, $Y = X^2$ between $X = 0$ and $X = 2$.

$$\text{Area} = \int_0^2 X^2 dX = \left[\frac{X^3}{3}\right]_0^2 = \left(\frac{8}{3} - 0\right) = 2\frac{2}{3} \quad \text{Square Units}$$

The quantity in the brackets $[\frac{X^3}{3}]_0^2$ is evaluated as:

$$\left[\frac{X^3}{3}\right] \text{ evaluated at } x = 2 - \left[\frac{X^3}{3}\right] \text{ evaluated at } x = 0$$

The integral can be used to solve circuit problems as follows:

Example 6–4:

Given: A certain capacitor has zero charge. A current of $I = 5t^2$ is applied to the capacitor.

Find: The amount of charge in coulombs on the capacitor at 4.5 seconds.

$$Q = \int I dt = \int 5t^2 dt = 5\int t^2 dt = 5\,\frac{t^3}{3}$$

$$\text{At 4.5 seconds } Q = 5\,\frac{(4.5)^3}{3} = \frac{5 \times 91.125}{3}$$

$Q = 151.875$ coulombs

Example 6–5:

Given: A voltage of $V = 3t$ is applied to an inductor of 0.01 henries.

Find: The current through the inductor at $t = 0.5$ seconds.

$$I = \frac{1}{L}\int V dt$$

$$I = \frac{1}{.01}\int 3t dt$$

$$I = 300 \int t dt = 150t^2$$

At $t = 0.5 \longrightarrow I = 150 \times (0.5)^2 = \underline{37.5 \text{ amps}}$

Constants of Integration

Up to this point, $\int X^N \, dX$ has been equal to:

$$\frac{X^{N+1}}{N+1} \quad \text{Since} \quad \frac{d}{dX}\frac{X^{N+1}}{N+1} = X^N$$

However, $\frac{X^{N+1}}{N+1}$ is not the only function whose derivative is X^n. $\frac{X^{N+1}}{N+1} + K$ where K is an arbitrary constant, is the general solution to $\int X^N dX$. K is called the constant of integration. The value of K depends on the initial or boundary conditions of a given problem.

Example 6–6: The charge Q on a device is given by:

$$\boxed{Q = \int I dt}$$

If: $I = 3t$
And: The device had an initial charge of 5 coulombs
Find: An equation for Q

$$Q = \int 3tdt = \frac{3}{2}t^2 + K$$

$$Q = \frac{3}{2}t^2 + K$$

Since $Q = 5$ at $t = 0$,

Substitute these values in the equation to determine a value for K

$$5 = 0 + K \longrightarrow K = 5, \text{ so}$$

$$\boxed{Q = \frac{3}{2}t^2 + 5}$$

SECTION 2: HOW TO USE DERIVATIVES AND INTEGRALS TO SOLVE CIRCUIT PROBLEMS

There are several examples in electronics in which the variables are related in terms of derivatives and integrals. Table 6-1 lists some of these.

1. Charge Q and current I	$I = \frac{dQ}{dt}$	$Q = \int I dt$
2. Voltage and current in a capacitive circuit	$I = C\frac{dV}{dt}$	$V = \frac{1}{C}\int I dt$
3. Voltage and current in an inductive circuit	$V = -L\frac{dI}{dt}$	$I = \frac{1}{L}\int V dt$

Table 6-1

These relationships hold for sinusoidal voltages and currents, as well as for any other defined functions.

Example 6-7:

Let: V = a Sin (wt)

Find: I

$$I = c\frac{dV}{dt} \qquad \frac{dV}{dt} = \text{wa cos (wt)*}$$

$$I = \text{cwa cos (wt)}$$

$$I = \text{(cwa)} \times \text{(cos wt)}$$

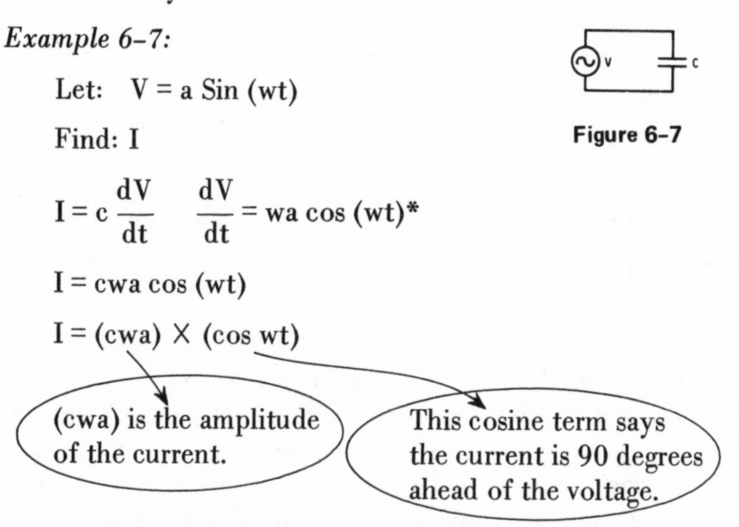

Figure 6-7

*Differential and integral operations for trigonometric functions are shown in the Table in Section 3.

The same results can be obtained using capacitive reactance.

$$I = \frac{E}{X_c} = \frac{a}{\frac{1}{wc}} = awc$$

Example 6–8:

Let: $V = a \sin(wt)$

Find: I

$$I = \frac{1}{L}\int V dt = \frac{a}{L}\int \sin wt\, dt$$

$$I = -\frac{a}{wL}\cos wt$$

$$I = \left(\frac{a}{wL}\right) \times (-\cos wt)$$

$\frac{a}{wL}$ is the amplitude

$-\cos wt$ says the current lags voltage by 90 degrees

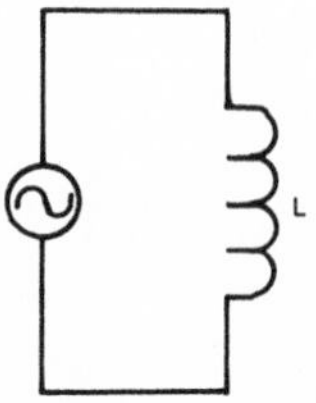

Figure 6–8

The same results can be obtained using inductive reactance.

$$I = \frac{E}{X_L} = \frac{a}{wL}$$

The above examples show that calculus can produce the same results as the use of inductive and capacitive reactance formulae, when the inputs are sinewaves. The real power of calculus is the ability to solve problems when the inputs are not sinewaves. Consider Examples 3 and 4 where the input is not sinusoidal.

Example 6–9: (See Figure 6–9.)

$V = 4t^2 - 2t + 3$

Find: I at t = 4.5 seconds

$$I = C\frac{dV}{dt} \qquad \frac{dV}{dt} = 8t - 2$$

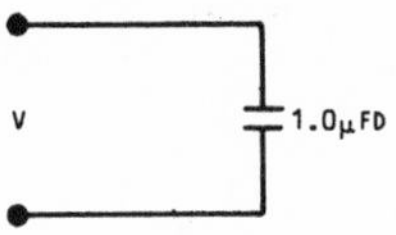

Figure 6–9

$I = (1 \times 10^{-6}) \times (8t - 2)$ at: $t = 4.5$

$I = (1 \times 10^{-6}) \times (8 \times 4.5 - 2) = 34 \times 10^{-6}$

$= 34$ microamps

Example 6-10: (See Figure 6-10.)

The Current $(I) = 3t^2$

Find: V at $t = 3.0$ seconds

$V = -L \dfrac{dI}{dt} \qquad \dfrac{dI}{dt} = 6t$

$V = (-2.5)(6 \times 3) = -45$ volts

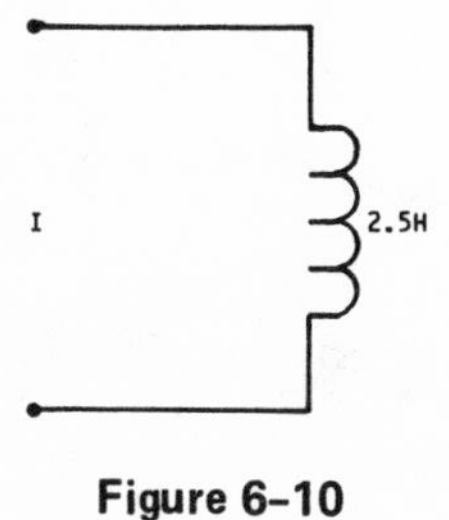

Figure 6-10

(The minus sign indicates the induced voltage is opposing the source voltage.)

SECTION 3: QUICK REFERENCE TABLES OF DERIVATIVES AND INTEGRALS

The following tables provide quick reference for most of the rules and specific functions you may encounter in solving electronic problems.

Derivatives

1. $\dfrac{d}{dX}$ constant $= 0$

2. $\dfrac{d}{dX}(u + V) = \dfrac{du}{dX} + \dfrac{dV}{dX}$

3. $\dfrac{d}{dX}uV = u\dfrac{dV}{dX} + V\dfrac{du}{dX}$

4. $\dfrac{d}{dX}\dfrac{u}{V} = \dfrac{V\dfrac{du}{dX} - u\dfrac{dV}{dX}}{V^2}$

5. $\dfrac{d}{dX}X^N = NX^{N-1}$

6. $\frac{d}{dX} K^X = K^X \log_e K$

7. $\frac{d}{dX} e^X = e^X$

8. $\frac{d}{dX} \sin u = \cos u \frac{du}{dX}$

9. $\frac{d}{dX} \cos u = -\sin u \frac{du}{dX}$

10. $\frac{d}{dX} \tan u = \sec^2 u \frac{du}{dX}$

11. $\frac{d}{dX} \cot u = -\csc^2 u \frac{du}{dX}$

Integrals

1. $\int dX = X + C$
2. $\int (u + V)dX = \int udX + \int VdX$
3. $\int adX = a \int dX$
4. $\int X^N dX = \frac{X^{N+1}}{N+1} + C$ (except for N = -1)
5. $\int \frac{dX}{X} = \log X + C$
6. $\int e^X dX = e^X + C$
7. $\int e^{aX} dX = \frac{e^{aX}}{a}$
8. $\int \sin XdX = -\cos X + C$
9. $\int \cos XdX = \sin X + C$
10. $\int \tan XdX = -\log(\cos X) + C$

11. $\int \cot X dX = \log(\sin X)$

12. $\int \sin^2 X dX = \frac{1}{2}X - \frac{1}{4}\sin 2X$

13. $\int \cos^2 X dX = \frac{1}{2}X + \frac{1}{4}\sin 2X$

14. $\int \frac{\sin X}{X} dX = X - \frac{X^3}{3.3!} + \frac{X^5}{5.5!} - \frac{X^7}{7.7!} + \ldots.$

15. $\int \log X dX = X \log X - X$

SECTION 4: USING HIGHER ORDER DERIVATIVES AND INTEGRALS

Derivatives

Consider the equation shown in Example 6-11. The derivative of this equation is shown in Step 1. This derivative is also an equation and a derivative can be found for it. This is shown in Step 2. We can continue taking derivatives until Step 4, when the derivative is zero.

Example 6-11: $Y = 3X^3 + 2X^2$ ⟶ Equation

Step 1 $\frac{dY}{dX} = 9X^2 + 4X$ ⟶ First Derivative

Step 2 $\frac{d^2Y}{dX^2} = 18X + 4$ ⟶ Second Derivative

Step 3 $\frac{d^3Y}{dX^3} = 18$ ⟶ Third Derivative

Step 4 $\frac{d^4Y}{dX^4} = 0$ ⟶ Fourth Derivative

Integrals

In an analogous manner, integrals can be continued as shown in Example 6-12.

Example 6-12: $Y = 2$

Step 1 $\int 2dX = 2X + K$

Step 2 $\int (2X + K)dX = \int 2XdX + \int KdX = X^2 + KX + K_1$

Step 3 $\int (X^2 + KX + K_1)dX = \int X^2 dX + \int KXdX + \int K_1 dX$

$$= \frac{X^3}{3} + K\frac{X^2}{2} + K_1 X + K_2$$

This example could be continued indefinitely. The symbols used to represent multiple integration are shown in Figure 6-11.

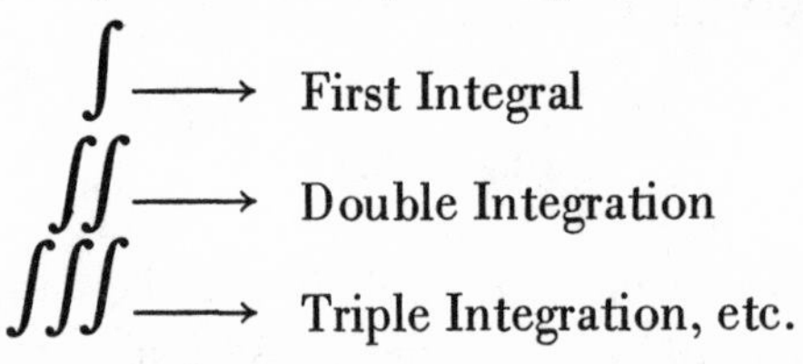

Figure 6-11

Example 6-13:

Assume that the charge (Q) moving through an inductor of 1.0 henries is given by the equation:

$Q = 3t^4 + t$

Find the voltage across the inductor at T = 5 seconds.
Solution: The charge is related to the current by $I = \frac{dQ}{dt}$ and the current to the voltage by $V = -L\frac{dI}{dt}$.

$$I = \frac{dQ}{dt} \longrightarrow \frac{dI}{dt} = \frac{d^2Q}{dt^2}$$

So: $V = -L\frac{d^2Q}{dt^2}$

$Q = 3t^4 + t$

$$\frac{dQ}{dt} = 12t^3 + 1$$

$$\frac{d^2Q}{dt^2} = 36t^2$$

$$V = -L\frac{d^2Q}{dt^2}$$

$$V = -L\,(36t^2)$$

$$t = 5.0 \qquad L = 1.0$$

$$V = -1.0\,(36 \times 25)$$

$$V = -900 \text{ volts}$$

SECTION 5: EXAMPLES OF CIRCUIT SOLUTIONS USING DIFFERENTIAL EQUATIONS

A differential equation is simply an equality containing derivatives and/or integrals. An example of a differential equation is shown in Figure 6-12. The "solution" of a differential equation is an equation. The arbitrary constants that are the result of integration can be determined by the boundary conditions of the problem.

$$\frac{d^2Y}{dX^2} = 2 \longrightarrow \text{Differential equation}$$

$$\frac{dY}{dX} = \int 2dX = 2X + K$$

$$Y = \int\int 2dX = X^2 + KX + K_1$$

Figure 6-12

Example 6-14: The current flowing into a 2.0 MFD capacitor is given by the equation $I = \sqrt{t}$. At $t = 0$, the charge on the capacitor is zero (boundary conditions).

Find the charge on the capacitor at t = 10 seconds.
Solution:

$$I = \frac{dQ}{dt}$$

$$\text{So: } \frac{dQ}{dt} = \sqrt{t} = t^{\frac{1}{2}}$$

$$Q = \int t^{\frac{1}{2}} dt = \frac{2t^{\frac{3}{2}}}{3} + K$$

The K constant of integration is to be determined by boundary conditions.

$$Q = \frac{2t^{\frac{3}{2}}}{3} + K$$

When t = 0, Q = 0, substitute these values in the equation:

$$0 = \frac{2(0)^{\frac{3}{2}}}{3} + K$$

$$0 = 0 + K \qquad \text{So: } K = 0$$

$$Q = \frac{2t^{\frac{3}{2}}}{3} = \frac{2\sqrt{t^3}}{3}$$

At: t = 10 seconds

$$Q = \frac{2\sqrt{10^3}}{3} = \frac{2\sqrt{1000}}{3} = \frac{2 \times 31.62}{3}$$

$$Q = 21.08 \text{ coulombs}$$

The time constant curves involved in charging a capacitor are derived by the use of differential equations. Consider the circuit in Figure 6-13.

If the capacitor has no charge at the time t = 0, and the switch is closed at t = 0, how does the voltage across the capacitor change with time?

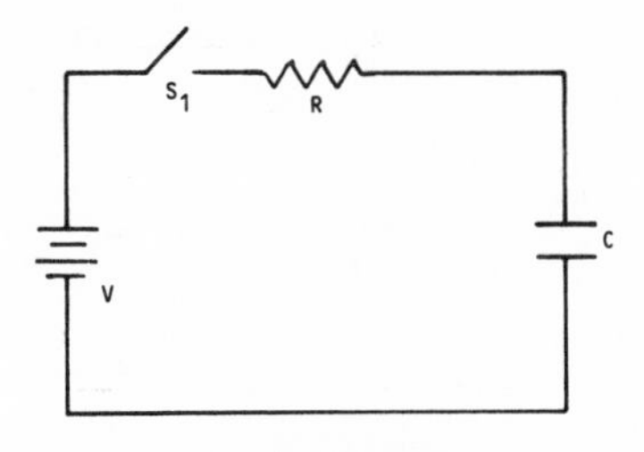

Figure 6-13

Kirchoff's Law says that at any time, the voltage drop across R, plus the voltage across C, must equal the source voltage.

So: $V_C + V_R = V$

$$V_C = \frac{1}{C} \int Idt \qquad V_R = IR$$

$$\frac{1}{C} \int Idt + IR - V = 0$$

Differentiating:

$$\frac{1}{C} I + R \frac{dI}{dt} = 0$$

or

$$R \frac{dI}{dt} + \frac{1}{C} I = 0$$

$$\frac{dI}{dt} = -\frac{1}{RC} I$$

The solution to this differential equation must involve an e^X term, since the equation says that the derivative $\frac{dI}{dt}$ and the current I are equal, and the function e^X satisfies this condition (see tables in Section 3). The solution to this differential equation is:

$$I = \frac{V}{R} e^{-t/RC}$$

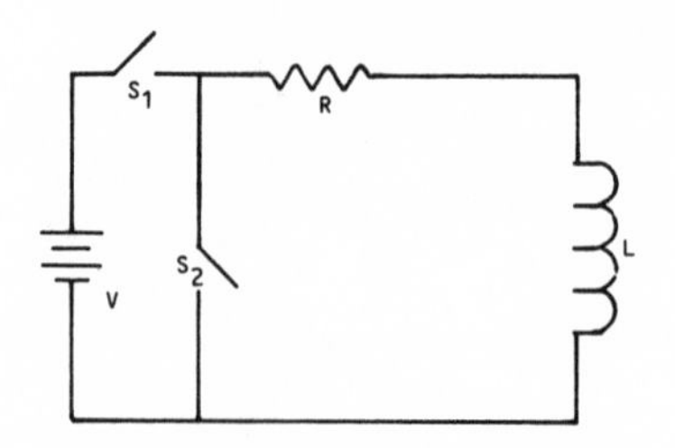

Figure 6-14

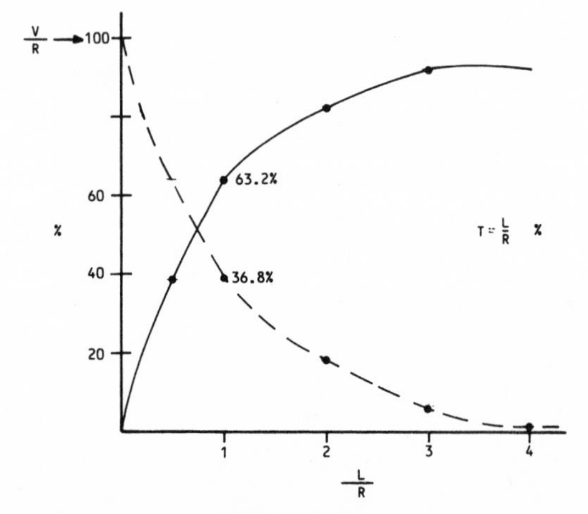

Figure 6-15A

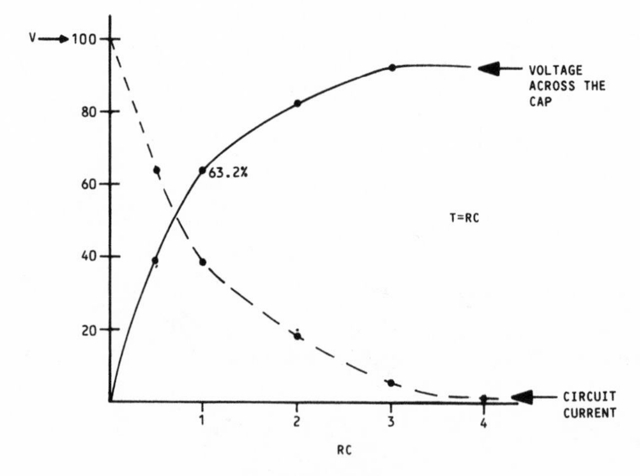

Figure 6-15B

At $t = 0$, all the voltage is dropped across the resistor and the current in the circuit is equal to $\frac{V}{R}$. When (t) is very large, the current ceases to flow and all the voltage is across the capacitor. A similar thing occurs in a series RL circuit. Kirchoff's Laws applied to the circuit in Figure 6-14. Assume S_1 has been closed for a long period of time so that the circuit current is steady. At the time $t = 0$, S_1 is opened and S_2 closed. The magnetic field around L will collapse and current will flow through L_1, R_1, and S_2. By Kirchoff:

$$-L\frac{dI}{dt} = RI$$

$$\text{or} \quad L\frac{dI}{dt} + RI = 0$$

The solution to this differential equation has the form e^X the same as a capacitive circuit. The exact solution is:

$$I = \frac{V}{R}e^{\frac{-R}{L}t}$$

The exponential charging of RC and RL circuits can be shown graphically as in Figure 6-15.

Infinite series provide a way to determine the value of a function such as e^x or $\tan^{-1} x$ to as close an approximation as needed. They reduce such functions to simplified algebraic functions. To determine e^x for any value of x, simply look for the expansion series in the quick reference tables and calculate the value.

A very useful example of an infinite series is the Fourier series. Given a periodic wave form, the Fourier series can be used to determine the amplitude of each harmonic present in the waveform. This can apply to square waves, triangle waves and distorted sinewaves.

Bessel functions are included in this chapter because of their role in explaining frequency modulation. If you are involved with F.M., whether it be broadcasting or two-way radio, the section on Bessel functions will be very helpful.

DEFINITION OF MACLAURIN AND TAYLOR SERIES

The MacLaurin method uses the function and its derivatives evaluated at $F(X) = 0$ to give an algebraic series for approximating the function. For values of X close to zero, the MacLaurin series converges rapidly. This means that it takes very few terms to obtain a very close approximation.

The Taylor series is very similar to the MacLaurin series, but it is more useful than MacLaurin for values of X removed from 0. The MacLaurin series takes many terms to produce a close approximation when X is much greater than zero. The Taylor series converges rapidly for values of X close to an arbitrary value (a).

MacLaurin Series

The MacLaurin series is found from the *General Formula . . .*

$$f(X) = f(0) + f'(0)X + \frac{f''(0)}{2!}X^2 + \ldots . \frac{f^N(0)}{N!}X^N \ *$$

f(0) → Function evaluated at zero

$\frac{f''(0)}{2!}X^2$ → Functions second derivative evaluated at zero divided by 2 factorial times X^2.

Example 1:

Find the MacLaurin series for $F(X) = e^x$

$f(X) = e^x$ $f(0) = 1$

$f'(X) = e^x$ $f'(0) = 1$

$f''(X) = e^x$ $f''(0) = 1$

$f^N(X) = e^x$ $f^N(0) = 1$

Substituting these into the general formula yields:

$f(0)$	$+ f'(0)X$	$+\frac{f''(0)}{2!}X$	$+\frac{f'''(0)}{3!}X^3$
$= 1$	$+ X$	$+\frac{X^2}{2}$	$+\frac{X^3}{6}$

---

*The Notation used has the following meaning:

f(0) ⟶ the value of the function when the variable is zero.

f'(0) ⟶ the ' indicates the first derivative. This is then the value of the first derivative evaluated at zero.

$\frac{f''(0)}{2!}$ ⟶ f''(0) indicates the second derivative evaluated at zero. The ! sign is a factorial. For example:
4! is 4 factorial
$4! = 4 \cdot 3 \cdot 2 \cdot 1 = 24$

Therefore: $e^x = 1 + X + \frac{X^2}{2} + \frac{X^3}{6} + \ldots \frac{X^N}{N!}$

This series can be used to determine $F(X) = e^x$ when X is close to zero.

Example 1: Determine e^x when $x = 0.1$.

$$e^x = 1 + X + \frac{X^2}{2} + \frac{X^3}{6}$$

$$= 1 + 0.1 + \frac{(0.1)^2}{2} + \frac{(0.1)^3}{6}$$

$$= 1 + 0.1 + .005 + .0001667$$

$$= 1.1051667$$

With $x = 0.1$, the MacLaurin series gave an extremely close approximation using only four terms.

Example 2: Determine sin X when X = 0.1 (X is in radians)

$f(X) = \sin X$ $\quad$ $f(0) = 0$

$f'(X) = \cos X$ $\quad$ $f'(0) = 1$

$f''(X) = -\sin X$ $\quad$ $f''(0) = 0$

$f'''(X) = -\cos X$ $\quad$ $f'''(0) = -1$

$f^4(X) = \sin X$ $\quad$ $f^4(0) = 0$

$f^5(X) = \cos X$ $\quad$ $f^5(0) = 1$

Substituting these values into the general formula:

$$\sin(X) = 0 + X + 0 - \frac{X^3}{3!} + 0 + \frac{X^5}{5!}$$

For X = 0.1

$$\sin(X) = 0.1 - \frac{(0.1)^3}{3!} + \frac{(0.1)^5}{5!}$$

$$\sin(X) = 0.1 - 0.0001667 + 0.0000000833$$

$$\sin(X) = 0.0998333833$$

Taylor Series

The Taylor series is found from the general formula.

$$f(X) = f(a) + f'(a)(X - a) + \frac{f''(a)}{2!}(X - a)^2 + \ldots \frac{f^N(a)}{N!}(X - a)^N$$

When the value of $(X - a)$ is made small, the Taylor series converges rapidly.

Example: Find the Taylor series for X^3.

$f(X) = X^3$ $\quad$ $f(a) = a^3$

$f'(X) = 3X^2$ $\quad$ $f'(a) = 3a^2$

$f''(X) = 6X$ $\quad$ $f''(a) = 6a$

$f'''(X) = 6$ $\quad$ $f'''(a) = 6$

$f^N(X) = 0$ $\quad$ $f^N(a) = 0$

$$f(X) = a^3 + 3a^2(X - a) + \frac{6a}{2!}(X - a)^2 + \frac{6}{3!}(X - a)^3$$

This expansion can be used to determine X^3 with any value of X.

Example 1: Find X^3 when $X = 2.97$

let $a = 3$ $\quad$ therefore $(X - a) = (-0.03)$

$$f(X) = 27 + 27(-0.03) + \frac{18}{2}(-0.03)^2 + \frac{6}{6}(-0.03)^3$$

$$= 27 - 0.81 + 0.0081 - 0.000027$$

$$f(X) = 26.198073$$

Example 2: Determine $\sqrt{3.9}$ from a series expansion.

$f(X) = X^{\frac{1}{2}}$ $\quad$ $f(a) = a^{\frac{1}{2}}$ $\quad$ let $a = 4$, therefore $(X - a) = (-0.1)$

$f'(X) = \frac{1}{2}X^{-\frac{1}{2}}$ $\quad$ $f'(a) = \frac{1}{2\sqrt{a}}$ $\quad$ $f(X) = \sqrt{a} + \frac{1}{2\sqrt{a}}(X - a) -$

$$- \frac{1}{4\sqrt{a^3}}(X - a)^2$$

$$f''(X) = -\frac{1}{4}X^{-\frac{3}{2}} \quad f''(a) = \frac{1}{4\sqrt{a^3}} \quad f(X) = 2 + \frac{1}{4}(-0.1) - \frac{1}{32}(.01)$$

$$f(x) = 1.974687$$

QUICK REFERENCE TABLES

Expontential:

$$e = 1 + \frac{1}{1} + \frac{1}{2!} + \frac{1}{3!} + \dots \frac{1}{N!}$$

$$e^x = 1 + X + \frac{X^2}{2!} + \frac{X^3}{3!} + \dots \frac{X^N}{N!}$$

$$a^x = 1 + X \log_e a + \frac{(X \log_e a)^2}{2!} + \dots \frac{(X \log_e a)^N}{N!}$$

$$\log_e X = (X - 1) - \frac{1}{2}(X - 1)^2 + \frac{1}{3}(X - 1)^3 - \frac{1}{4}(X - 1)^4 + \dots$$

For $(2 > X > 0)$

Trigonometric:

$$\text{Sin } X = X - \frac{X^3}{3!} + \frac{X^5}{5!} - \frac{X^7}{7!} + \dots$$

$$\cos X = 1 - \frac{X^2}{2!} + \frac{X^4}{4!} - \frac{X^6}{6!} + \dots$$

$$\sin^{-1} X = X + \frac{X^3}{6} + \left(\frac{1}{2} \cdot \frac{3}{4}\right)\frac{X^5}{5} + \left(\frac{1}{2} \cdot \frac{3}{4} \cdot \frac{5}{6}\right)\frac{X^7}{7} + \dots$$

For $(X^2 < 1)$

$$\tan^{-1} X = X - \frac{1}{3}X^3 + \frac{1}{5}X^5 - \frac{1}{7}X^7 + \dots$$

$(X^2 < 1)$

$$= \frac{\pi}{2} - \frac{1}{X} + \frac{1}{3X^3} - \frac{1}{5X^5} + \dots$$

$(X^2 > 1)$

DEFINITION OF THE FOURIER SERIES

Any periodic waveform, from a square wave to a triangle wave to whatever, is made up of sinewaves. (see Figure 7-1.) These sinewaves are harmonically related. Their frequencies are whole number multiples of a fundamental frequency. The Fourier series is a mathematical method determining the amplitude and phase relationship of each harmonic in the waveform. The amplitude of the harmonics must decrease as the frequency increases.

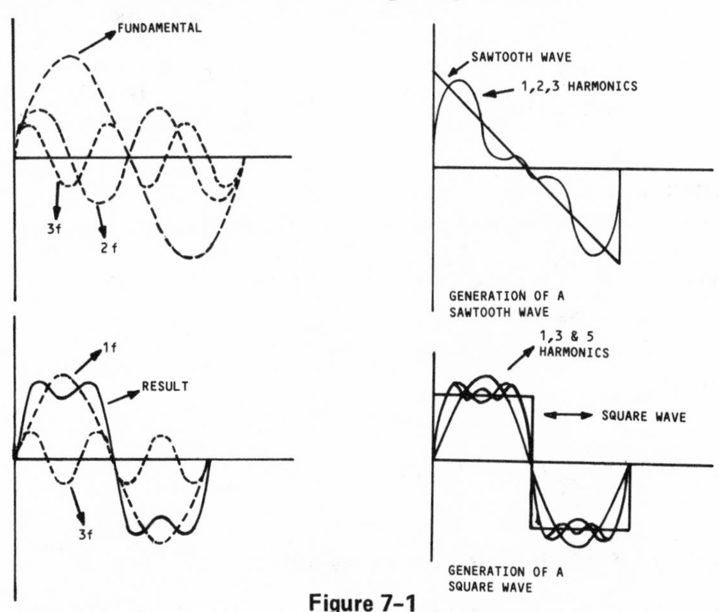

Figure 7-1

The Fourier series may be represented mathematically in several ways. All of these methods, however, are equivalent.

Angle Variable

$$f(t) = \frac{A_0}{2} + \sum_{N=1}^{\infty} (A_N \cos N\theta + B_N \sin N\theta)$$

Where:

$\frac{A_0}{2}$ = Average (D.C.) value over one period of the waveform.

$$A_N = \frac{1}{\pi}\int_{-\pi}^{\pi} f(\theta)\cos N\theta d\theta \qquad B_N = \frac{1}{\pi}\int_{-\pi}^{\pi} f(\theta)\sin N\theta d\theta$$

Time Variable

$$f(t) = \frac{A_0}{2} + \sum_{N=1}^{\infty} (A_N \cos Nwt + B_N \sin Nwt)$$

Where:

$$A_N = \frac{2}{T}\int_{-T/2}^{T/2} f(t)\cos Nwtdt \qquad B_N = \frac{2}{T}\int_{-T/2}^{T/2} f(t)\sin Nwtdt$$

Exponential Form

$$f(t) = \alpha_0 + \sum_{N=-\infty}^{\infty} \alpha_N e^{iNwt}$$

Where:

$$\alpha_N = \frac{1}{T}\int_{-T/2}^{T/2} f(t)\, e^{-iNwt}\, dt$$

How to Use Fourier on Periodic Waveforms

The following is an example of how to apply the formulas to a waveform to determine the harmonic content.

Example:

Find the Fourier series for the squarewave in Figure 7-2.

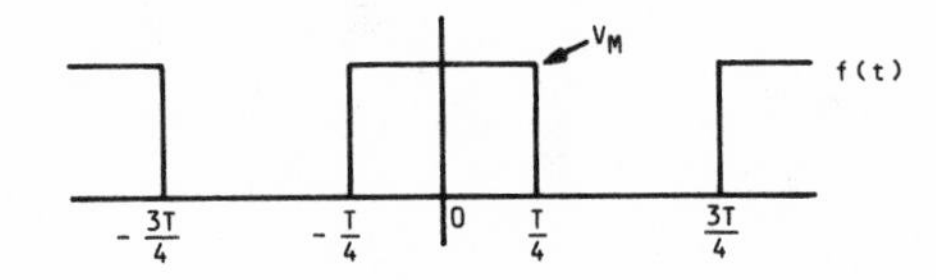

Figure 7-2

A square wave where:

For 1 period

$$F(T) = \begin{cases} 0 \text{ for } & -\frac{T}{2} < t < -\frac{T}{4} \\ V_M \text{ for } & -\frac{T}{4} < t < \frac{T}{4} \\ 0 \text{ for } & \frac{T}{4} < t < \frac{T}{2} \end{cases}$$

$$A_N = \frac{2}{T}\int_{-T/4}^{T/4} V_M \cos Nwtdt$$

$$= \frac{2V_M}{Nwt} \sin Nwt \Big|_{-T/4}^{T/4}$$

$$= \frac{2V_M}{N\pi} \sin\left(\frac{N\pi}{2}\right)$$

If $N = 2,4,6,8 \ldots A_N = 0 \longrightarrow$ no even harmonics

If $N = 1,5,9,13 \ldots A_N = \frac{2V_M}{N\pi}$

If $N = 3,7,11,15 \ldots A_N = -\frac{2V_M}{N\pi}$

$$B_N = \frac{2}{T}\int_{-T/4}^{+T/4} V_M \sin Nwtdt$$

$$B_N = -\frac{2V_M}{Nwt} \cos Nwt \Big|_{-T/4}^{T/4}$$

$B_N = 0 \longrightarrow$ no sine terms

$$f(t) = \frac{V_M}{2} + \frac{2V_M}{\pi}\left[\cos wt - \frac{1}{3}\cos 3wt + \ldots\right]$$

Ao/2 can be found from inspection. The average (D.C.) value is Vm/2.

The graph of harmonic amplitude is shown in Figure 7-3. The example used is symmetrical about the vertical axis. It

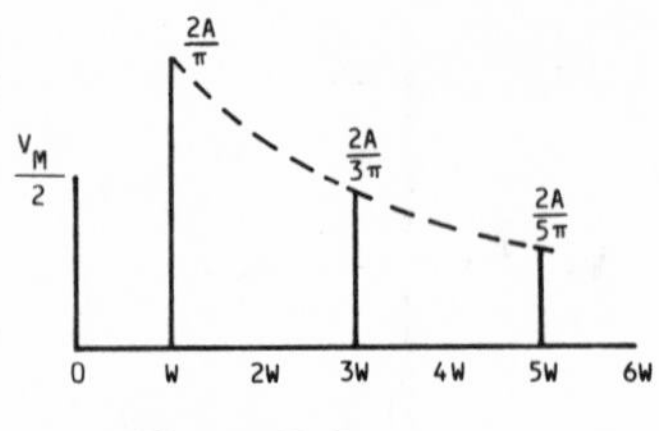

Figure 7-3

therefore contains only cosine terms. The Fourier series can often be simplified by conditions of symmetry as shown in Table 7-1.

Conditions	Property	A_N	B_N
$f(t) = f(-t)$	Cos only	$\frac{4}{T}\int_0^{T/2} f(t) \cos\frac{2\pi N}{T}tdt$	0
$f(t) = -f(-t)$	Sin only	0	$\frac{4}{T}\int_0^{T/2} f(t) \sin\frac{2\pi N}{T}tdt$
$f\left(t \pm \frac{T}{2}\right) = f(t)$	Even N only	$\frac{4}{T}\int_0^{T/2} f(t) \cos\frac{2\pi N}{T}tdt$	$\frac{4}{T}\int_0^{T/2} f(t) \sin\frac{2\pi N}{T}tdt$
$f\left(t \pm \frac{T}{2}\right) = -f(t)$	Odd N only	$\frac{4}{T}\int_0^{T/2} f(t) \cos\frac{2\pi N}{T}tdt$	$\frac{4}{T}\int_0^{T/2} f(t) \sin\frac{2\pi N}{T}tdt$

Table 7-1

Example: Find the Fourier series for the waveform in Figure 7-4.

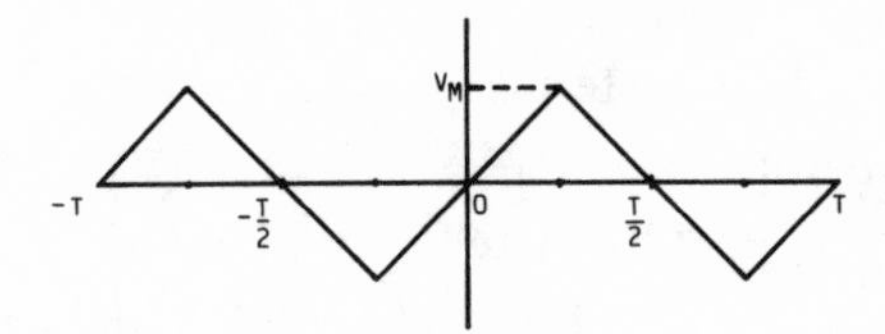

Figure 7-4

since: $f(t) = -f(-t) \longrightarrow A_N = 0$

$f\left(t \pm \frac{T}{2}\right) = -f(t) \longrightarrow$ odd harmonics only by inspection $\frac{Ao}{2} = 0 \longrightarrow$ no D.C. component

$$f(t) = \frac{4Vm}{T}t \quad \text{for} \quad -\frac{T}{4} \leq t \leq \frac{T}{4}$$

$$B_N = \frac{4}{T}\int_{-T/4}^{T/4} \frac{4Vm}{T}t \sin\left(\frac{2\pi N}{T}\right)tdt$$

$$B_N = \frac{8}{T}\int_{0}^{T/4} \frac{4Vm}{T}t \sin\left(\frac{2\pi N}{T}\right)tdt$$

$$B_N = \frac{8Vm}{N^2\pi^2} \sin \frac{N\pi}{2} \Bigg\} \quad \text{For } N = 0,2,4 \ldots . \sin \frac{N\pi}{2} = 0$$

$$B_N = \frac{8Vm}{N^2\pi^2} \quad \text{For } N = 1,5,9 \ldots$$

$$B_N = -\frac{8Vm}{N^2\pi^2} \quad \text{For } N = 3,7,11 \ldots$$

$$f(t) = \frac{8Vm}{\pi^2}\left(\sin wt - \frac{1}{9} \sin 3wt + \frac{1}{25} \sin 5wt \ldots .\right)$$

Some common waveforms with their Fourier series are given in the following illustrations in Figures 7-5 to 7-14.[1]

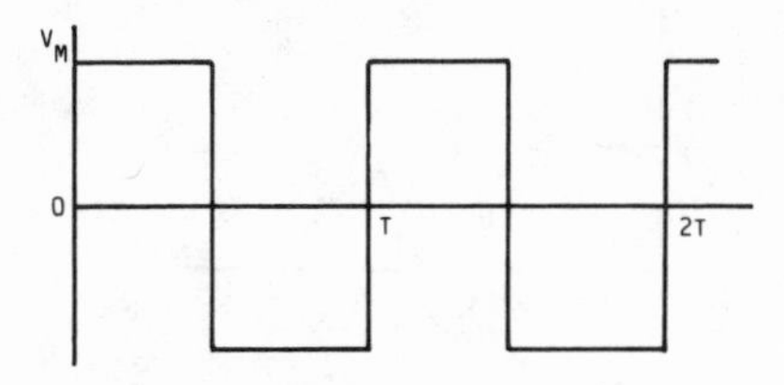

Figure 7-5

$$V = \frac{4Vm}{\pi}\left[\sin wt + \frac{1}{3} \sin 3wt + \frac{1}{5} \sin 5wt + \ldots\right]$$

---

[1]E. Norman Lurch, *Electric Circuit Fundamentals*, © 1979, pp. 520–526. Reprinted by permission of Prentice-Hall, Inc., Englewood Cliffs, N.J.

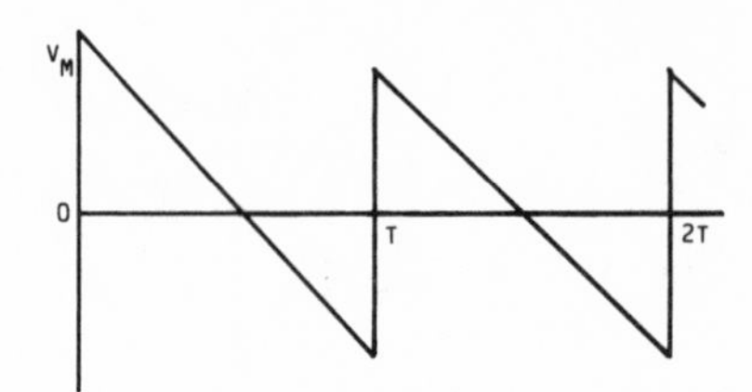

Figure 7-6

$$V = \frac{2Vm}{\pi}\left[\sin wt + \frac{1}{2}\sin 2wt + \frac{1}{3}\sin 3wt + \ldots.\right]$$

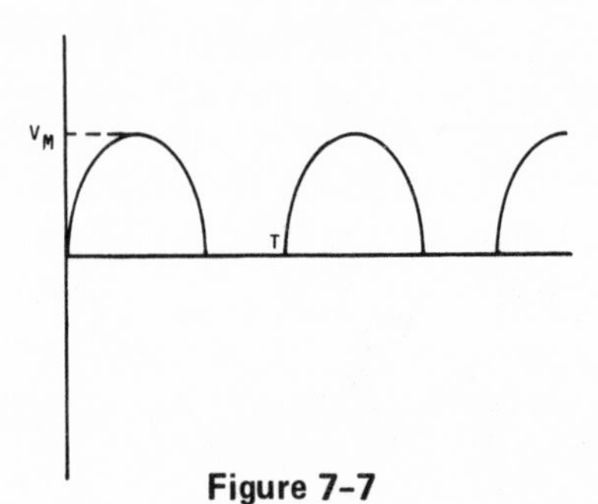

Figure 7-7

$$V = \frac{Vm}{\pi} + \frac{Vm}{2}\sin wt - \frac{2Vm}{\pi}\sum_{N=2,4,6\ldots}^{\infty}\frac{\cos Nwt}{N^2 - 1}$$

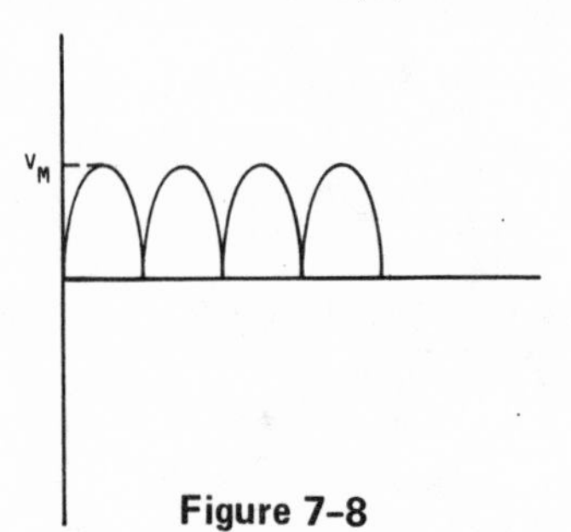

Figure 7-8

$$V = \frac{2Vm}{\pi} - \frac{4Vm}{\pi}\sum_{N=2,4,6\ldots}^{\infty}\frac{\cos Nwt}{N^2 - 1}$$

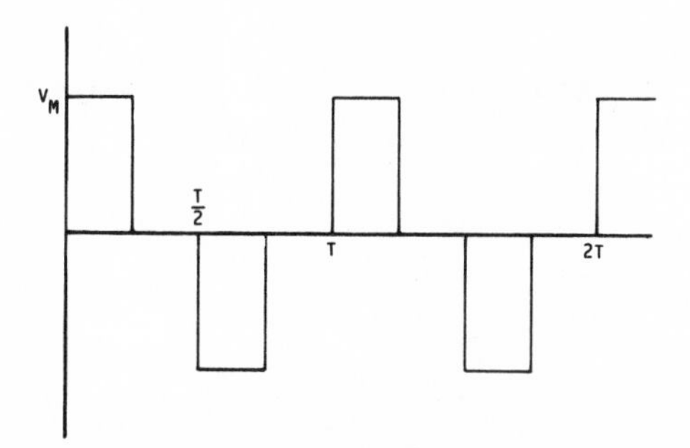

Figure 7-9

$$V = \frac{2\sqrt{2}}{\pi} Vm \left[\cos wt + \frac{1}{3} \cos 3wt - \frac{1}{5} \cos 5wt - \frac{1}{7} \cos 7wt + \ldots \right]$$

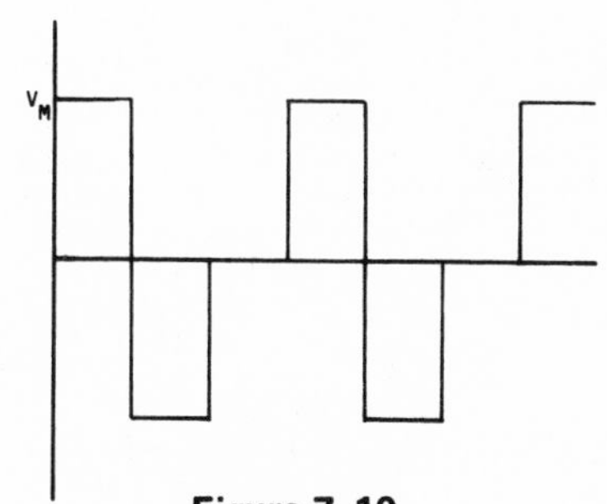

Figure 7-10

$$V = \frac{2Vm}{\pi} \left[\sin wt - \frac{2 \sin 2wt}{2} + \frac{3 \sin 3wt}{3} + \frac{\sin 5wt}{5} - \frac{2 \sin 6wt}{6} + \frac{\sin 7wt}{7} + \ldots \right]$$

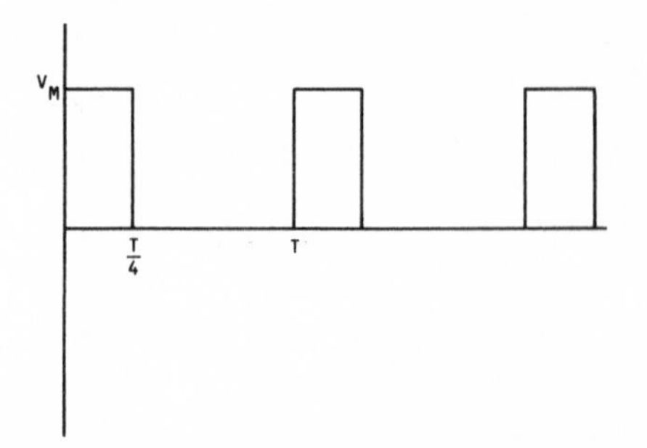

Figure 7-11

$$V = \frac{1}{4} Vm + \frac{\pi}{2} Vm\left(\frac{\sqrt{2}}{2} \cos wt + \frac{1}{2} \cos 2wt + \frac{\sqrt{2}}{6} \cos 3wt - \frac{\sqrt{2}}{10} \cos 5wt + \ldots \right)$$

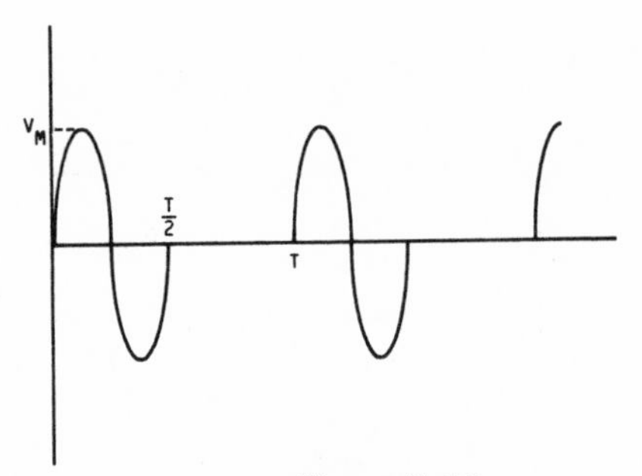

Figure 7-12

$$V = \frac{1}{2} Vm \sin 2wt + \frac{4}{\pi} Vm \left(\frac{\sin wt}{3} + \frac{\sin 3wt}{5} - \frac{\sin 5wt}{21} - \ldots \right)$$

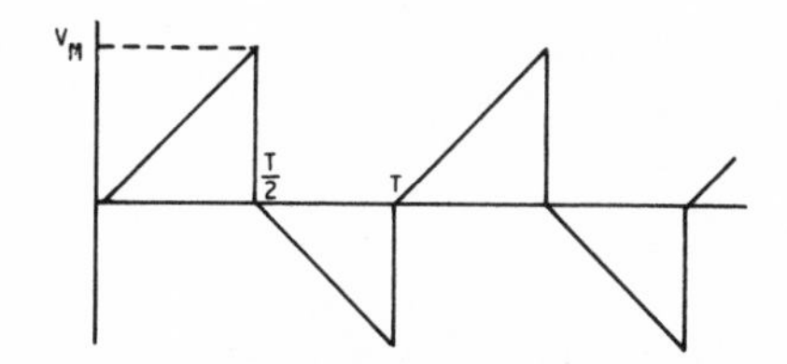

Figure 7-13

$$V = \frac{4}{\pi^2} Vm \left(\cos wt + \frac{\cos 3wt}{9} + \frac{\cos 5wt}{25} + \ldots\right)$$

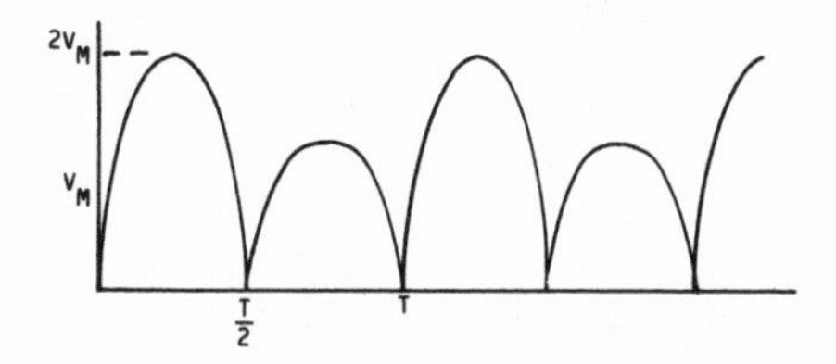

Figure 7-14

$$V = \frac{3}{\pi} Vm + \frac{1}{2} Vm \sin wt - \frac{6}{\pi} Vm \left(\frac{\cos 2wt}{3} + \frac{\cos 4wt}{15} + \frac{\cos 6wt}{35} + \ldots\right)$$

Example: In a half-wave rectified sinewave Figure 7-7, what is the amplitude of the tenth harmonic? The peak voltage of the half wave is 100 volts.

Solution: The Fourier series for a half-wave signal is shown in the examples. The tenth harmonic is:

$$\frac{2Vm}{\pi} \left(\frac{\cos 10wt}{100 - 1}\right)$$

The amplitude is $\dfrac{2Vm}{99\pi}$

$= 0.643$ volts

Definition of the Fourier Transform

The Fourier series is useful for analyzing periodic waveforms. It gives the discrete frequency components of the waveform. The Fourier series is limited to periodic waveforms. It cannot be used to analyze pulses of any shape if they are from a single pulse. To analyze single pulses for frequency content, it is necessary to use the Fourier transform.

Fourier Transform

The Fourier transform is defined by the following pair of equations.

$$f(t) = \frac{1}{2\pi} \int_{-\infty}^{+\infty} g(w)\, e^{iwt} dw$$

$$g(w) = \int_{-\infty}^{+\infty} f(t)\, e^{-iwt} dt$$

These equations allow functions in the time domain to be transformed into the frequency domain and from the frequency domain back into the time domain.

How to Use Fourier to Analyze Pulses

Example:

Determine the frequency content of the voltage pulse shown in Figure 7-15.

$$f(t) = \begin{matrix} Vm \;\; 0 < t < T \\ 0 \text{ elsewhere} \end{matrix}$$

$$g(w) = \int_{0}^{T} Vm\, e^{-iwt} dt$$

Figure 7-15

$$= \frac{Vm}{iw}(1 - e^{-iwt})$$

$$g(w) = \frac{2V_M}{w} \sin\left(\frac{wT}{2}\right) e^{-iwT/2}$$

$$\text{The amplitude of } g(w) = \frac{2V_M}{w} \sin\left(\frac{wT}{2}\right)$$

This can also be written:

$$V_M T \left(\frac{\sin\left(\frac{wT}{2}\right)}{\frac{wT}{2}} \right)$$

This gives a frequency distribution of:

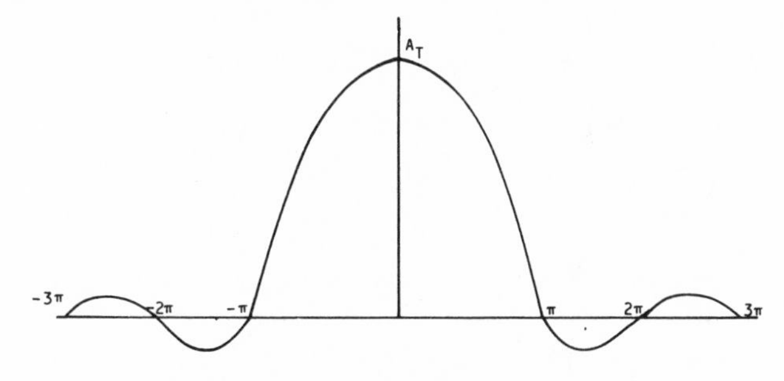

Figure 7-16

The frequency distribution of a single pulse is a continuous function, whereas the frequency distribution of a periodic waveform consists of discrete frequencies.

Understanding Bessel Functions

Bessel functions are encountered in electronics whenever the mathematical form used is [SIN (SIN θ)]. The most common example is the frequency modulation of a sine wave such as in F.M. broadcasting.

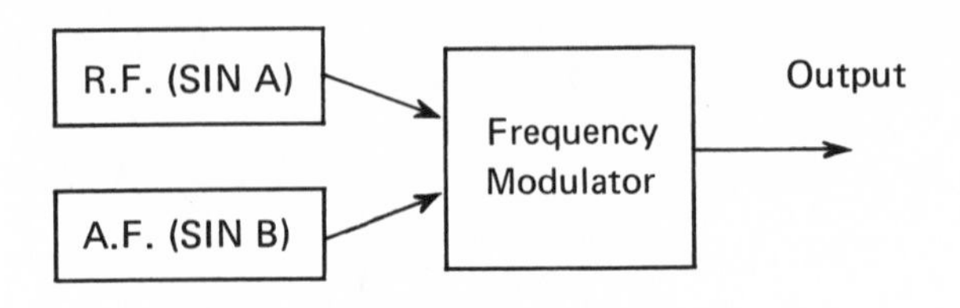

Figure 7-17

The output signal in Figure 7-17 is:

Output = SIN (A + M SIN B)

Where M is the modulation index.

When this expression is expanded using Fourier series, the following results are obtained:

$$\begin{aligned} \text{Output} = {} & J_0(M)\ \text{SIN A} \\ & + J_1(M)\ \text{SIN}\ (A + B) - \text{SIN}\ (A - B) \\ & + J_2(M)\ \text{SIN}\ (A + 2B) + \text{SIN}\ (A - 2B) \\ & + J_3(M)\ \text{SIN}\ (A + 3B) - \text{SIN}\ (A - 3B) \\ & + J_4(M)\ \text{SIN}\ (A + 4B) + \text{SIN}\ (A - 4B) \end{aligned}$$

The output is composed of a carrier J_0 (M) SIN A plus sidebands that are spaced at whole number multiples of the audio frequency away from the carrier.

The amplitudes of the carrier and sidebands are determined by the J_n (M) coefficients. These coefficients are known as Bessel functions of the first kind and of the order shown by the subscript. Bessel functions of the first kind are solutions to the differential equation:

$$M^2 \frac{D^2 Y}{DM^2} + M \frac{DY}{DM} + (M^2 - N^2)Y = 0$$

This solution is:

$$J_n(M) = \left(\frac{M}{2}\right)^N \left[\frac{1}{N!} - \frac{(M/2)^2}{1!(N+1)!} + \frac{(M/2)^4}{2!(N+2)!} + \ldots\ldots\right]$$

The Bessel function values are shown graphically in Figure

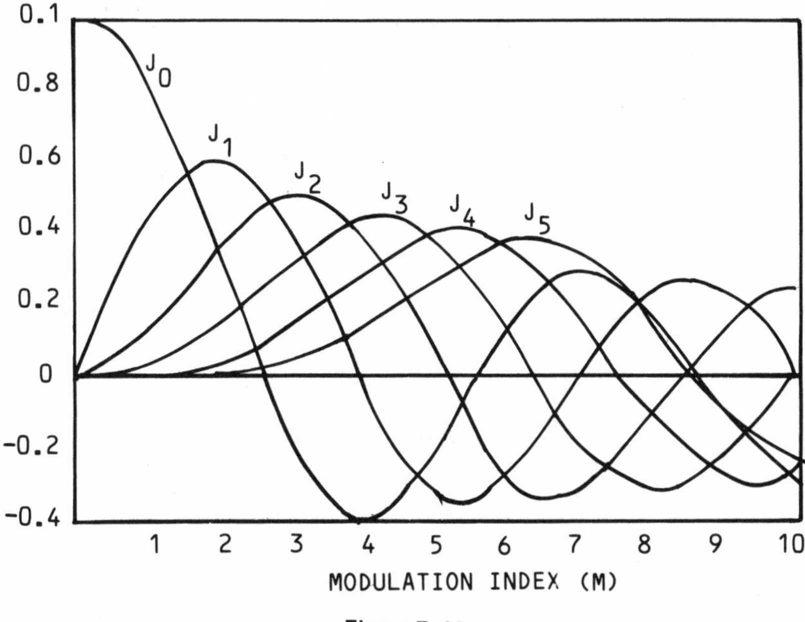

Figure 7-18

7-18. Note: there are certain modulation indexes where the carrier amplitude (J_0) disappears. These are (2.4), (5.6), (8.7), etc. The amplitude of the carrier and significant sidebands for various values of modulation index are shown in Table 7-2.

The fact that the carrier disappears for certain modulation indices gives an excellent method for calibration of the frequency deviation of an F.M. transmitter. (See Figure 7-19.)

Consider the following:

1. Modulation index = carrier deviation/audio frequency.
2. At modulation index ≈ 2.49 the carrier disappears.
3. Use a setup as shown in Figure 7-20.
4. Set the audio oscillator frequency to 1 KHZ and output to zero. The receiver should have a strong reading on the S-meter when tuned properly.
5. Increase the output amplitude of the audio oscillator until the S-meter reaches a minimum.

6. This represents a modulation index of 2.49, the first carrier null.
7. The transmitter must have a frequency deviation of 2.49 KHZ since (Audio frequency) × (mod index) = carrier deviation.

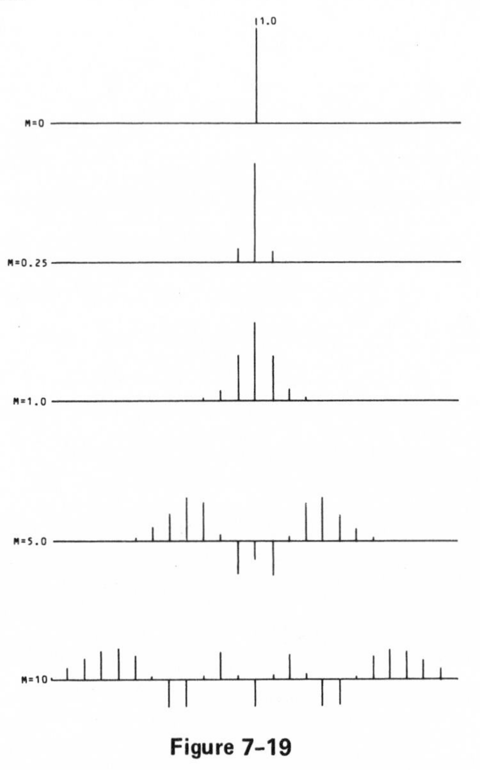

Figure 7-19

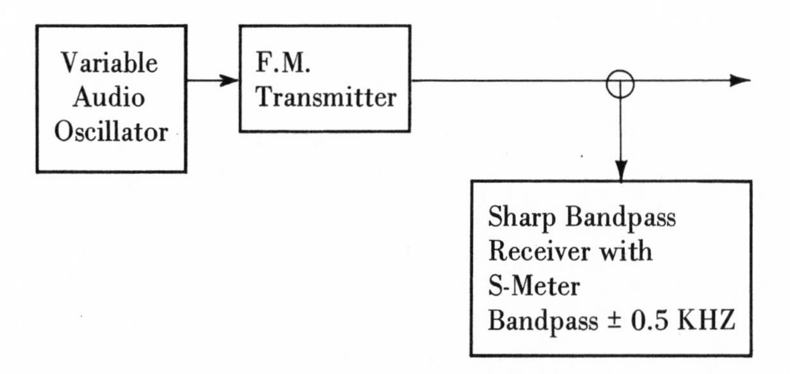

Figure 7-20

Mod Index	J_0	J_1	J_2	J_3	J_4	J_5	J_6	J_7	J_8	J_9	J_{10}	J_{11}	J_{12}
0.0	1.00												
0.25	0.98	0.12											
0.5	0.94	0.24	0.03										
1.0	0.77	0.44	0.11	0.02									
1.5	0.51	0.56	0.23	0.06	0.01								
2.0	0.22	0.58	0.35	0.13	0.03								
2.5	-0.05	0.50	0.45	0.22	0.07	0.02							
3.0	-0.26	0.34	0.49	0.31	0.13	0.04	0.01						
4.0	-0.40	-0.07	0.36	0.43	0.28	0.13	0.05	0.02					
5.0	-0.18	-0.33	0.05	0.36	0.39	0.26	0.13	0.05	0.02				
6.0	0.15	-0.28	-0.24	0.11	0.36	0.36	0.25	0.13	0.06	0.02			
7.0	0.30	0.00	-0.30	-0.17	0.16	0.35	0.34	0.23	0.13	0.06	0.02		
8.0	0.17	0.23	-0.11	-0.29	-0.10	0.19	0.34	0.32	0.22	0.13	0.06	0.03	
9.0	-0.09	0.24	0.14	-0.18	-0.27	-0.06	0.20	0.33	0.30	0.21	0.12	0.06	0.03
10.0	-0.25	0.04	0.25	0.06	-0.22	-0.23	-0.01	0.22	0.31	0.29	0.20	0.12	0.06
12.0	0.05	-0.22	-0.08	0.20	0.18	-0.07	-0.24	-0.17	0.05	0.23	0.30	0.27	0.20
15.0	-0.01	0.21	0.04	-0.19	-0.12	0.13	0.21	0.03	-0.17	-0.22	-0.09	0.10	0.24

Table 7-2

8 EXAMPLES OF SOLUTIONS WITH MONOGRAMS AND GRAPHS

Monograms and graphs provide a convenient and quick method for solving electronic problems. This chapter contains examples of solutions from Ohm's Law, to transmission line impedance, to coil winding problems. Quite often these graphical solutions are quicker than working through complicated formulas. In each case, a "blank" copy is provided for your use and examples of how to use the monogram/graph are also given.

SECTION 1: IMPEDANCE MONOGRAM

To use this monogram, simply connect the three vertical lines with a straight line. (See Figures 8-1 and 8-2.)

Example: If a 5KΩ resistor is connected in series with an 8 KΩX_c, what is the total impedance in ohms?

Solution: A straight line between X = 8 and R = 5 intersects the Z axis at 9.4. The total impedance is then 9400 ohms.

The monogram can be used for any power of ten. For example, the above example could have been 50Ω resistor, 80ΩX_c and 94ΩZ.

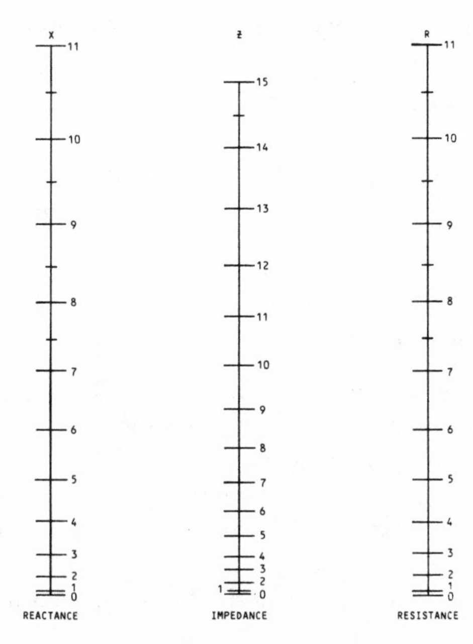

Figure 8–1

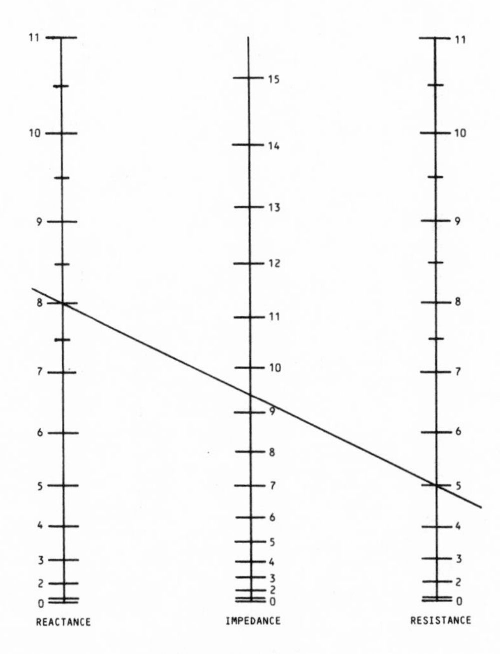

Figure 8–2

SECTION 2: COIL WINDING MONOGRAM

This monogram is more complex than most, since it requires the use of two straight lines instead of one. With a little practice however, it is quite easy. (See Figures 8-3 and 8-4.)

Example: Determine the inductance of the following coil.

Length = 2 inches.
Diameter = 2.0 inches
Number of turns = 150

Solution: The diameter/length ratio = 1.0.
Draw a straight line from the number of turns (150) to the diameter/length ratio (1.0).
From the point where this line crosses the (plot) axis, draw another straight line to the diameter (2.0). The point where this second line crosses the inductance axis is the desired inductance value, or 800μH.

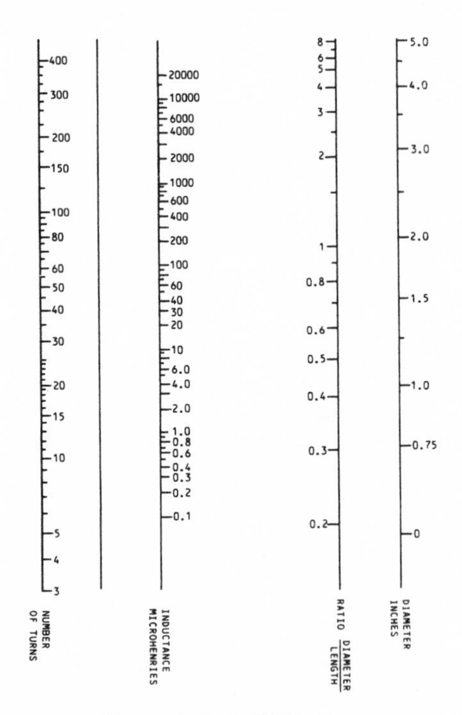

Figure 8-3. Coil Winding

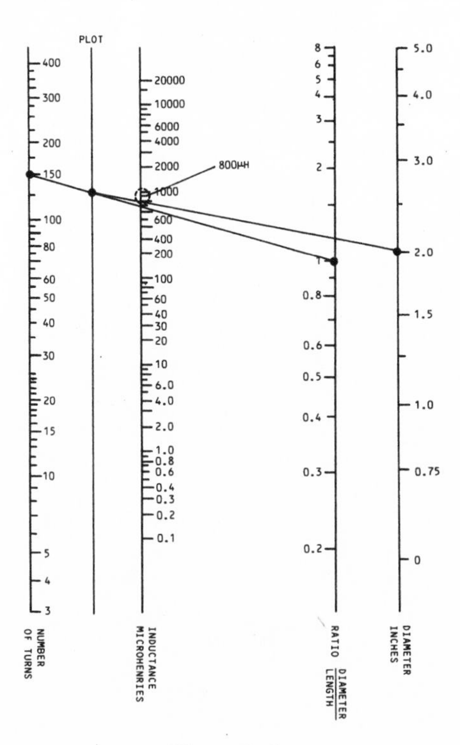

Figure 8-4

SECTION 3: OHM'S LAW MONOGRAM

To use this monogram, simply connect the three vertical lines with a straight line. (See Figures 8-5 and 8-6.)

Example: The line drawn can represent several different values.

1. 30 volts, 6 amps, 50 ohms
2. 30 volts, 6 ma, 5K ohms
3. 3 volts, 0.6 amps, 5 ohms
4. etc.

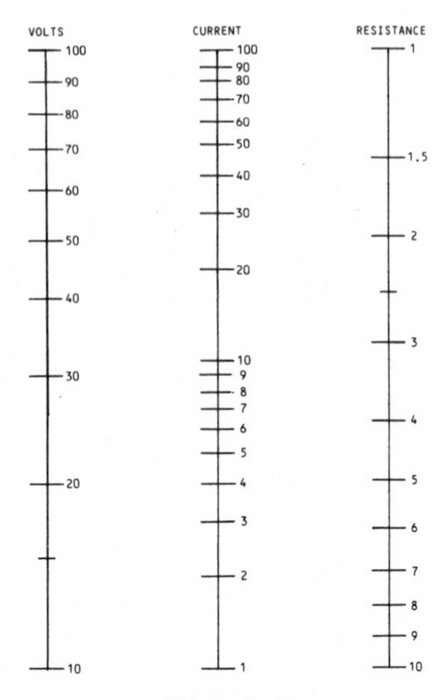

Figure 8-5

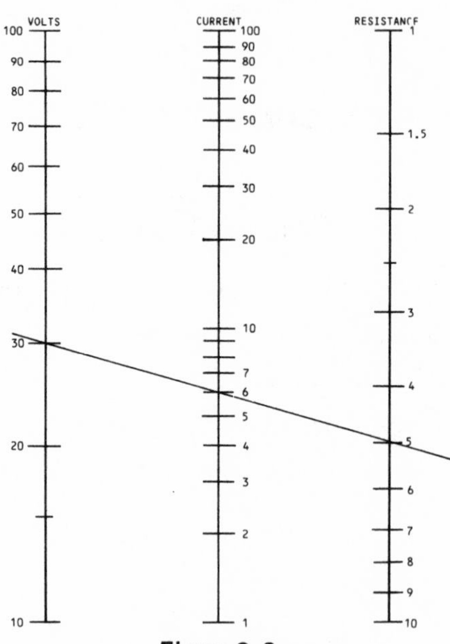

Figure 8-6

SECTION 4: RESONANCE MONOGRAM

This monogram is extremely useful. All of the following can be determined. (See Figure 8-7.)

A. Capacitive reactance

B. Inductive reactance

C. Resonant frequency for combinations of inductance and capacitance.

The examples in Figures 8-8 and 8-9 show that a 1mh inductor and a 10pf capacitor resonate at 1.592 mhz and, at this frequency, each has a reactance of 10KΩ.

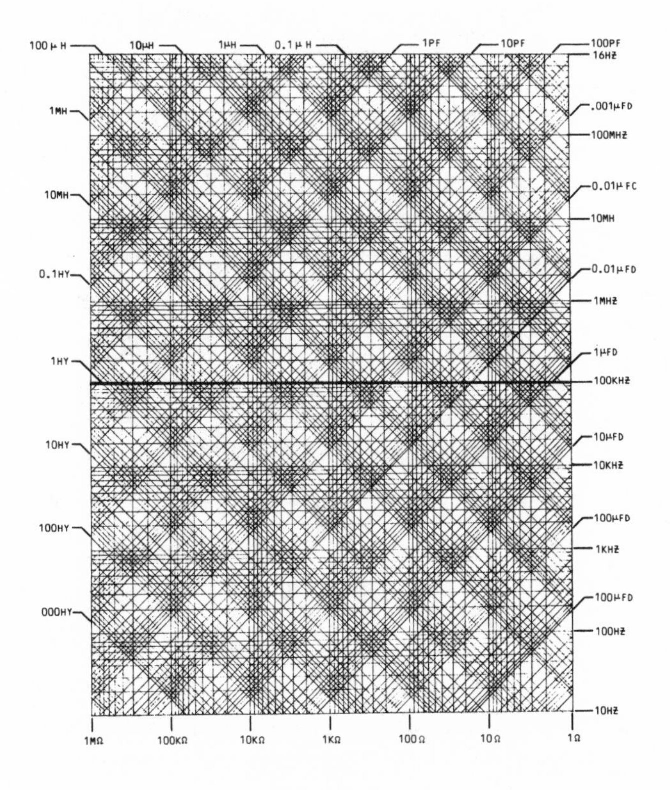

Figure 8-7

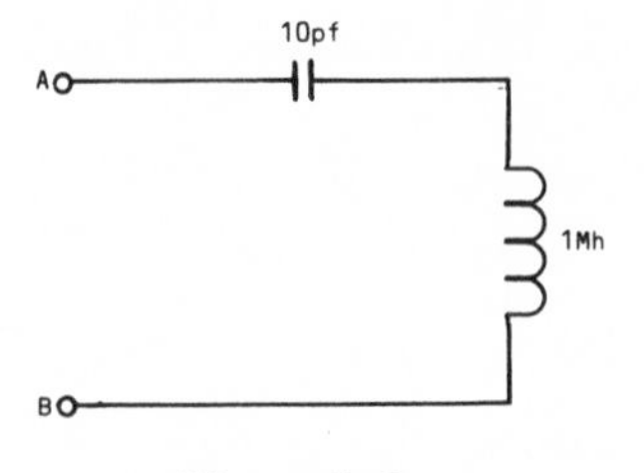

Figure 8-8

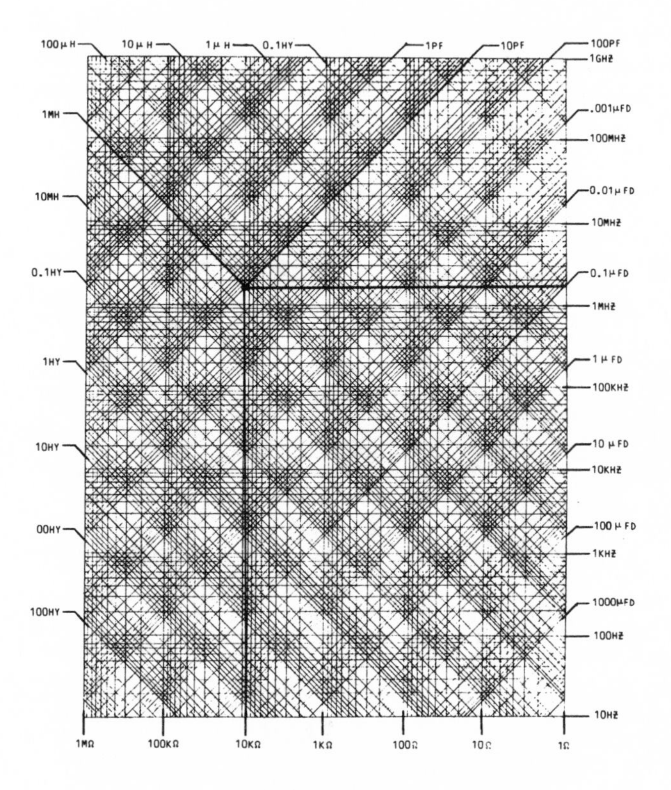

Figure 8-9

Example: 10 PF, 1 MH — Resonate at 1.592MHZ; $X_L = X_c = 10K\Omega$

SECTION 5: PARALLEL RESISTOR GRAPHS

Using formulas is one way to determine the total resistance of parallel resistors. Without the aid of a calculator, the parallel resistor formulas can be very cumbersome.

The following examples show how to solve parallel resistor circuits graphically. Simply draw two lines whose lengths are proportional to the resistance values. Connect the ends of the lines as shown in Figure 8-10. The point where the two lines intersect represents the value of total resistance.

When more than two resistors are in parallel, they can be treated as shown in Figure 8-11.

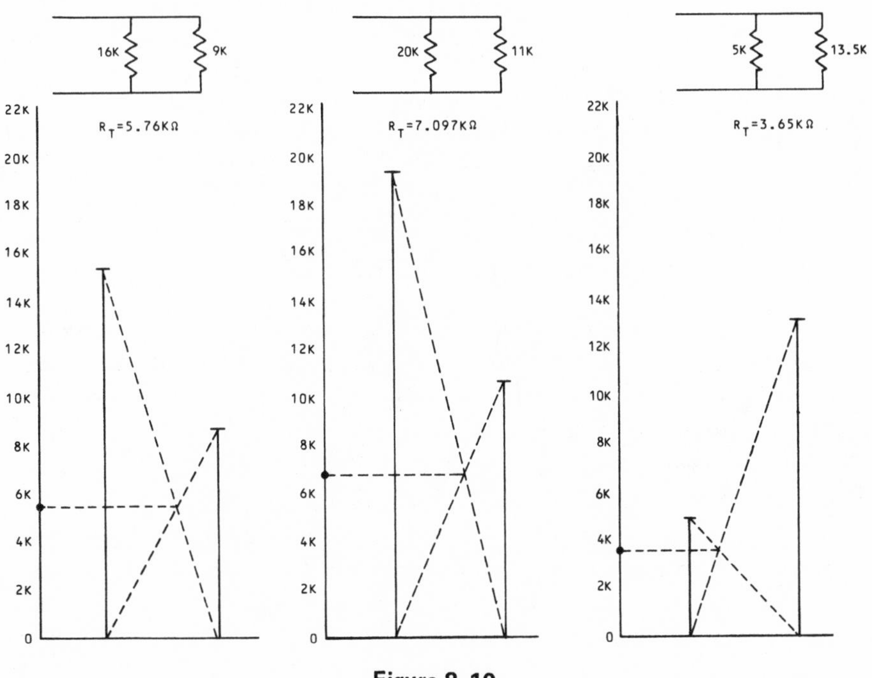

Figure 8-10

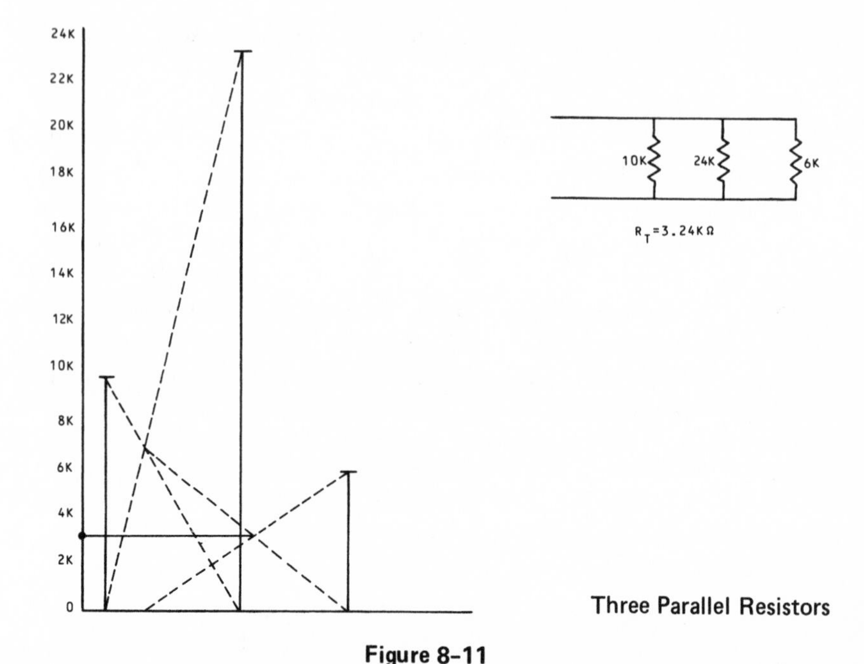

Figure 8–11

SECTION 6: LOAD LINE ANALYSIS

Load lines are powerful tools in the solution of circuit analysis problems. They can be used to solve linear circuits, but most important, they can be used to solve circuits with nonlinear devices such as diodes and transistors.

Construction of a load line is shown in Figure 8–12. The example uses a 1-ohm resistor as the load. However, any device may be used as a load, as long as the E-I curve is known. Figure 8–13 shows a load line analysis using an unknown load Z. In order to construct a load line, two points must be determined. These points are on the vertical and horizontal axes.

A. The point on the voltage axis is equal to the source voltage or the open circuit load voltage.

B. The point on the current axis is the current when the load is short-circuited, or $I = E/R$ where E is the source voltage and R is the series resistor.

C. The point of intersection of the load line and the device E-I curve is the point where the circuit operates.

D. A line drawn from the point of intersection, perpendicular to the current axis, intersects at the point representing the total circuit current.

E. A line drawn from the point of intersection, perpendicular to the voltage axis, intersects at a point showing the voltage division between the load and the series resistor.

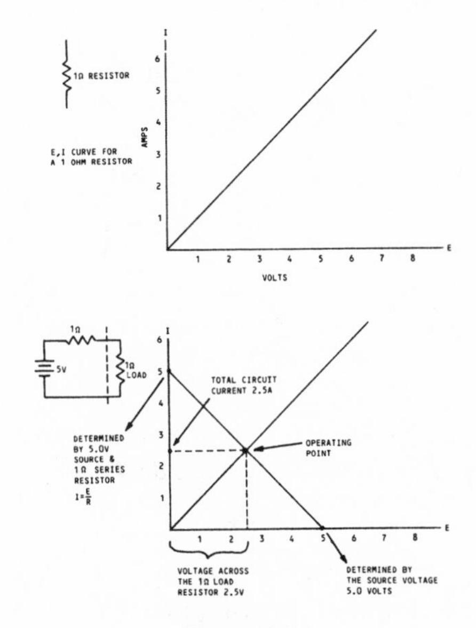

Figure 8–12

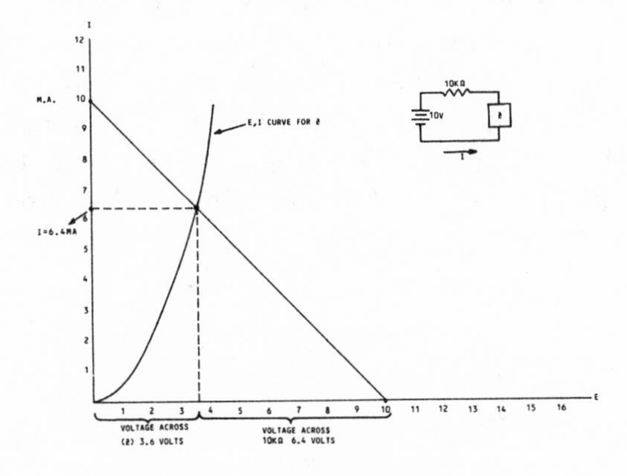

Figure 8–13

Figure 8–14 illustrates the effect of:

1. A change in the source voltage.
2. A change in the series resistor.

A change in the source voltage moves the load line parallel to itself, while a change in the series resistor changes the point of intersection with the current axis.

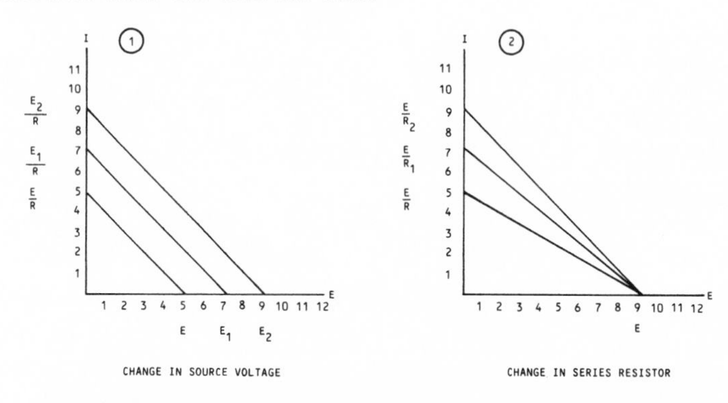

1. Changing the source voltage moves the load line in parallel.
2. Changing the series resistor changes the slope of the load line.

Figure 8–14

An excellent example of load line analysis is shown in Figure 8–15. With the circuit shown, Illustration 1 shows the voltage across the diode (0.11) volts. If we should increase the source voltage for an instant, as in Illustration 2, the voltage across the diode would increase to 9.78 volts. When the source voltage drops back to 1.0 volts, the load line is the same as Illustration 1, but now the operating point is different and the voltage across the diode is 0.68 volts. (See Illustration 3.) This load line analysis shows the basic operation of tunnel diodes when used as digital memory devices.

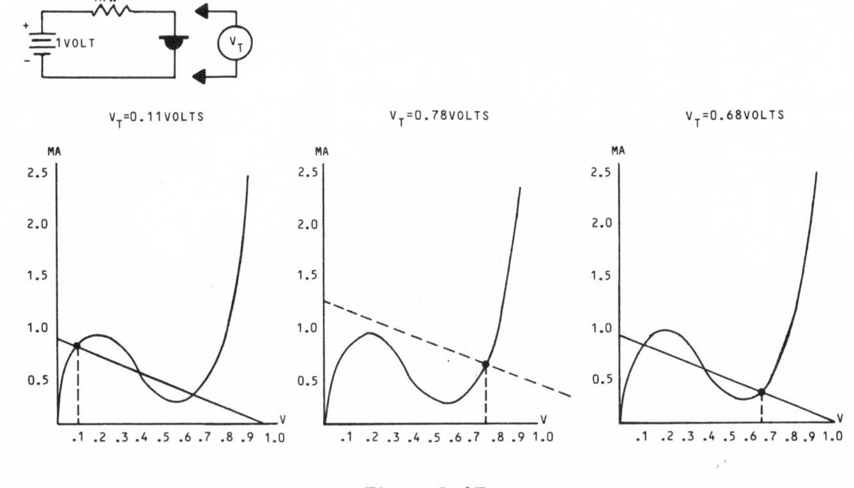

Figure 8–15

N-Channel J-Fet

A set of characteristic curves for an n-channel j-fet is shown in Figure 8–16. A load line has been drawn based on a supply voltage of +15 volts and a total series resistance of 1250 ohms.

A. The Bias voltage is the drop across R_s

$V_{rs} = .0075 \times 133 = 1.0$ volts

B. The voltage across R_1

$V_{rl} = 1117 \times .0075 = 8.4$ volts

C. The voltage on the drain of Q_1 with respect to ground is 6.16 volts.

Figure 8–17 shows graphically what happens when a 1-volt peak-to-peak signal is applied to the gate.

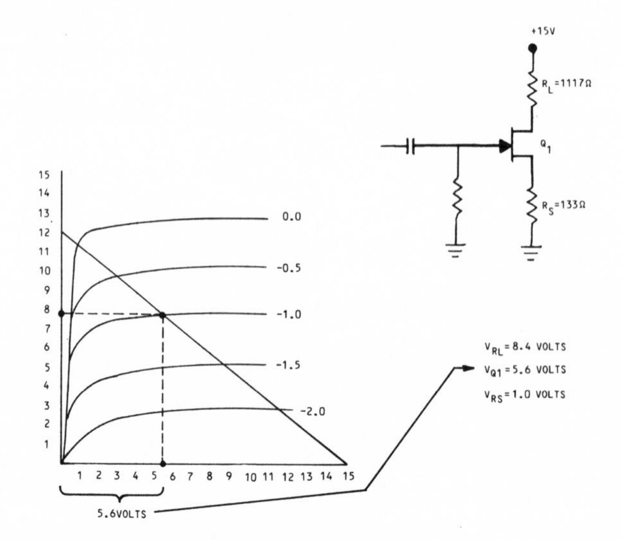

Figure 8-16

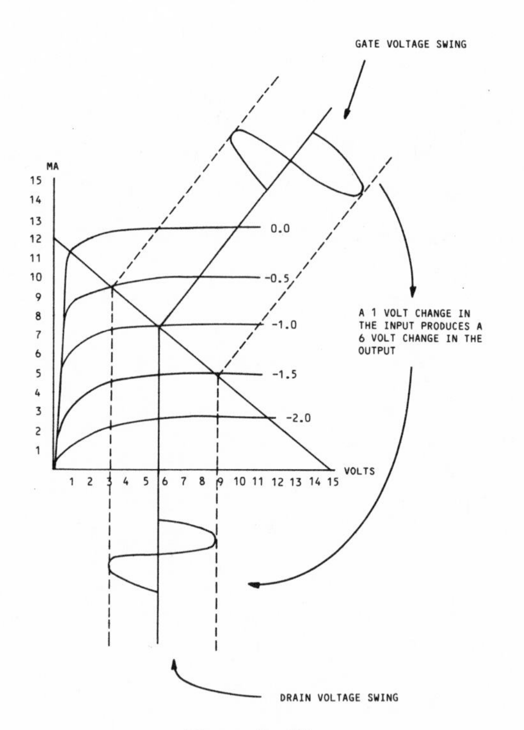

Figure 8-17

SECTION 7: SMITH CHARTS

The Smith chart is a graphical method of solving problems involving antennas and transmission lines. It is a very powerful tool and, when used properly, it can solve almost any problem involving transmission lines. Figure 8-18 shows the basic areas of the Smith chart. The use of the chart is best explained in an example.

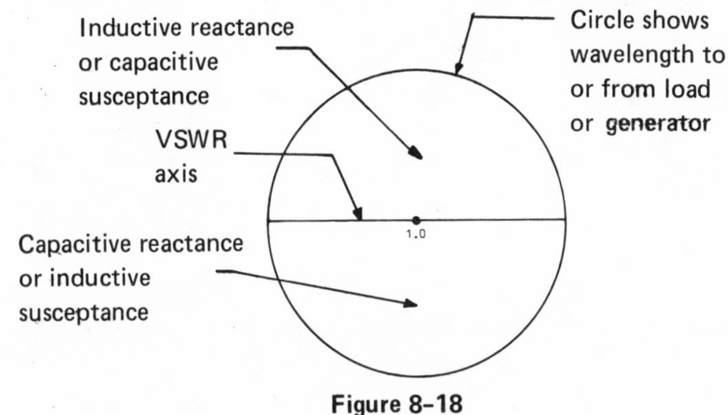

Figure 8-18

Example 1:

To demonstrate the use of the chart, consider a 300 ohm transmission line that has been terminated with a 900 ohm resistor. Since this is not a "matched" condition, the line will not be flat. There will be standing waves on the line, and the standing wave ratio will be the same as the impedance ratio (900/300 equals 3 to 1 VSWR).

To represent this condition on a Smith chart, a circle is drawn. The center of the circle is at 1.0 on the VSWR axis and the circumference passes through 3.0, since the VSWR is 3.0 to 1. This is shown in Figure 8-19, along with the starting point for this example. At the starting point, the impedance is purely resistive and is three times the transmission line impedance (900Ω load 300Ω line).

As we move down the transmission line, it is the same as moving along the VSWR circle in the clockwise direction. Point 1 represents a distance that is 0.1 λ* from the 900Ω load. Point 2 represents 0.2 λ from the load, etc.

---

*λ is the symbol used to indicate wavelength.

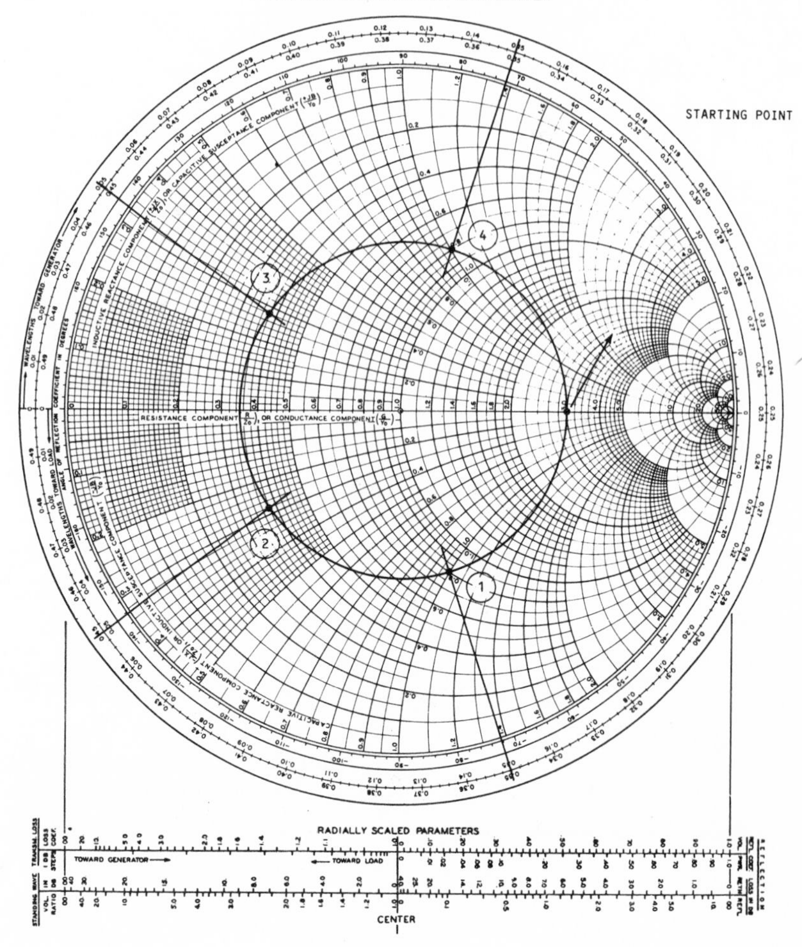

Figure 8-19

The impedance at each of these points is read from the coordinates of the graph.

Point 1 The graph reads (0.8 - 1.01J). The impedance is found by multiplying this by 300.
$Z = 300 \times (0.8 - 1.01J) = 240 - 303J$
Point 2 $Z = 300 \times (0.36 - 0.29J) = 108 - 87J$
Point 3 $Z = 300 \times (0.36 + 0.29J) = 108 + 87J$
Point 4 $Z = 300 \times (0.8 + 1.01J) = 240 + 303J$

These impedances are shown graphically in Figure 8-20. The equivalent series circuit is also shown. Note that one full circle on

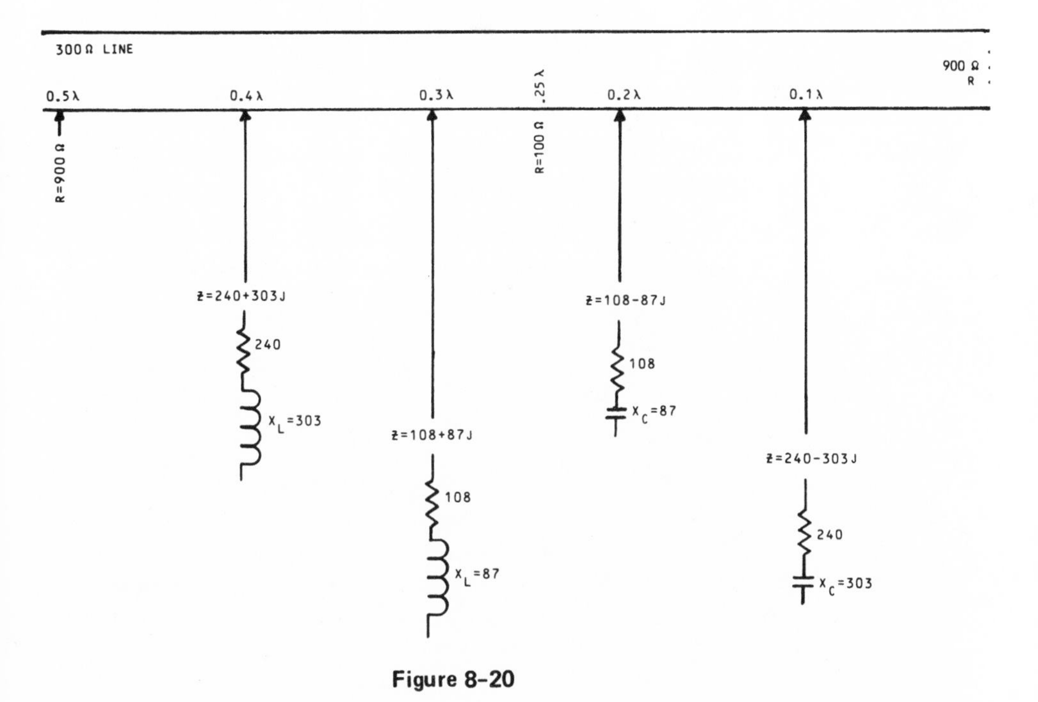

Figure 8-20

the Smith chart represents a half wavelength and that conditions along a transmission line repeat themselves every half wavelength.

Velocity Factor

When the terminology "wavelengths along a transmission line" is used, it implies that the velocity factor for that line has been taken into account. The wavelength along a line will always be shorter than a free space wavelength, according to the following formula:

$$\lambda(\text{meters}) = \left(\frac{300}{\text{Freq in mhz}}\right) \times (\text{velocity factor})$$

$$\text{Velocity factor} \leq 1.0$$

The velocity factor for various transmission lines is given in Table 8-1.

Type	Factor*
1. Open wire line	0.99 - 1.0
2. 300 Ribbon	0.82
3. 75 Ribbon	0.71
4. { Coaxial RG - 8,11,58,59 }	0.66

*These values are approximate and will vary with different manufacturers.

Table 8-1

Example 2:

A 50-ohm transmission line is terminated with an R.C. network, where Z = 150 - 60J. The velocity factor is 0.66. The frequency is 10 MHZ.

How far down the transmission line is the first point where the impedance is purely resistive, and what is the value of the resistance?

Step 1. Normalize the load impedance with respect to the line $(150 - 60J)/50 = (3 - 1.2J)$.

2. Find this point on the Smith Chart. (Figure 8-21, Point A.)
3. Draw a circle through Point A, with the center on the VSWR axis at Point 1.0.
4. From Point A, follow the circle clockwise until it intersects the VSWR axis (Point B).
5. At this point, the impedance is purely resistive. Its value is: $(0.28) \times (50) = 14$ ohms

   (0.28) ↓ graph; (50) ↓ line Z

6. The distance down the transmission line is found on the circumference of the chart $(0.50 - 0.27) = 0.23\lambda$.
7. $1.0\ \lambda$ on this line would be

$$\frac{300}{10} \times 0.66 = 19.8 \text{ meters}$$

$$0.23\lambda = 19.8 \times 0.23$$

$$= \boxed{4.554 \text{ meters}}$$

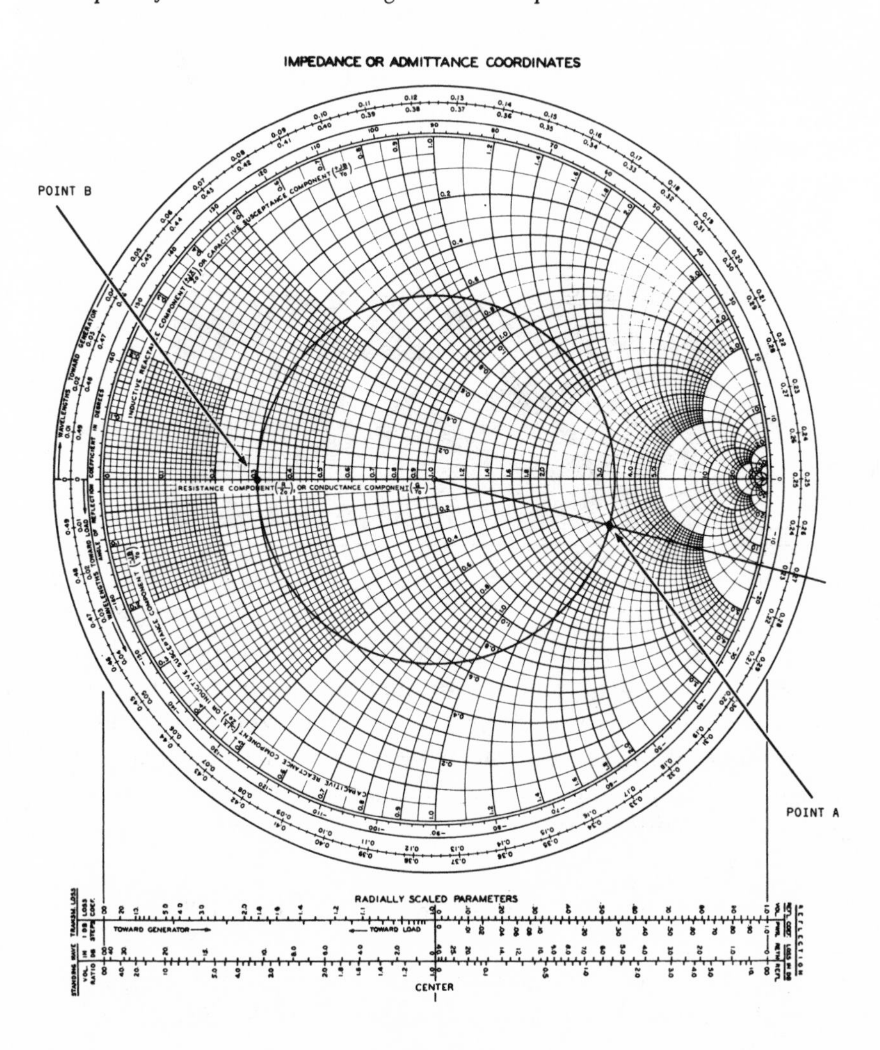
IMPEDANCE OR ADMITTANCE COORDINATES
POINT B
POINT A
RESISTANCE COMPONENT $\left(\frac{R}{Z_0}\right)$, OR CONDUCTANCE COMPONENT $\left(\frac{G}{Y_0}\right)$
INDUCTIVE REACTANCE COMPONENT $\left(\frac{+jX}{Z_0}\right)$, OR CAPACITIVE SUSCEPTANCE COMPONENT $\left(\frac{+jB}{Y_0}\right)$
CAPACITIVE REACTANCE COMPONENT $\left(\frac{-jX}{Z_0}\right)$, OR INDUCTIVE SUSCEPTANCE COMPONENT $\left(\frac{-jB}{Y_0}\right)$
WAVELENGTHS TOWARD GENERATOR
WAVELENGTHS TOWARD LOAD
ANGLE OF REFLECTION COEFFICIENT IN DEGREES
RADIALLY SCALED PARAMETERS
TOWARD GENERATOR
TOWARD LOAD
CENTER

Figure 8–21

This chapter contains reference tables for analog systems such as logarithmic and trigonometric functions, to digital information like ASCII and binary to decimal conversion. Whenever necessary, an example of how to use the table is provided. This chapter will provide an easy, time-saving reference source, available whenever you need this information.

SECTION 1: TRIGONOMETRIC FUNCTIONS

The trigonometric functions of sine, cosine, and tangent are presented for angles between 0 degrees and 90 degrees in one-degree steps. (See Table 9-1.)

0	sin	0.000	3	sin	0.052	6	sin	0.105
	cos	1.000		cos	0.999		cos	0.995
	tan	0.000		tan	0.052		tan	0.105
1	sin	0.017	4	sin	0.070	7	sin	0.122
	cos	1.000		cos	0.998		cos	0.993
	tan	0.017		tan	0.070		tan	0.123
2	sin	0.035	5	sin	0.087	8	sin	0.139
	cos	0.999		cos	0.996		cos	0.990
	tan	0.035		tan	0.087		tan	0.141

Table 9-1. Trigonometric Functions

Angle	Function	Value	Angle	Function	Value	Angle	Function	Value
9	sin	0.156	22	sin	0.375	35	sin	0.574
	cos	0.988		cos	0.927		cos	0.819
	tan	0.158		tan	0.404		tan	0.700
10	sin	0.174	23	sin	0.391	36	sin	0.588
	cos	0.985		cos	0.920		cos	0.809
	tan	0.176		tan	0.424		tan	0.727
11	sin	0.191	24	sin	0.407	37	sin	0.602
	cos	0.982		cos	0.914		cos	0.799
	tan	0.194		tan	0.445		tan	0.754
12	sin	0.208	25	sin	0.423	38	sin	0.616
	cos	0.978		cos	0.906		cos	0.788
	tan	0.213		tan	0.466		tan	0.781
13	sin	0.225	26	sin	0.438	39	sin	0.629
	cos	0.974		cos	0.899		cos	0.777
	tan	0.231		tan	0.488		tan	0.810
14	sin	0.242	27	sin	0.454	40	sin	0.643
	cos	0.970		cos	0.891		cos	0.766
	tan	0.249		tan	0.510		tan	0.839
15	sin	0.259	28	sin	0.469	41	sin	0.656
	cos	0.996		cos	0.883		cos	0.755
	tan	0.268		tan	0.523		tan	0.869
16	sin	0.276	29	sin	0.485	42	sin	0.669
	cos	0.961		cos	0.875		cos	0.743
	tan	0.287		tan	0.554		tan	0.900
17	sin	0.292	30	sin	0.500	43	sin	0.682
	cos	0.956		cos	0.866		cos	0.731
	tan	0.306		tan	0.577		tan	0.933
18	sin	0.309	31	sin	0.515	44	sin	0.695
	cos	0.951		cos	0.951		cos	0.719
	tan	0.325		tan	0.601		tan	0.966
19	sin	0.326	32	sin	0.530	45	sin	0.707
	cos	0.946		cos	0.848		cos	0.707
	tan	0.344		tan	0.625		tan	1.000
20	sin	0.342	33	sin	0.545	46	sin	0.719
	cos	0.940		cos	0.839		cos	0.695
	tan	0.364		tan	0.649		tan	1.04
21	sin	0.358	34	sin	0.559	47	sin	0.731
	cos	0.934		cos	0.829		cos	0.682
	tan	0.384		tan	0.675		tan	1.07

Table 9-1. Trigonometric Functions (contd.)

Angle	Function	Value	Angle	Function	Value	Angle	Function	Value
48	sin	0.743	61	sin	0.875	74	sin	0.961
	cos	0.669		cos	0.485		cos	0.276
	tan	1.11		tan	1.80		tan	3.490
49	sin	0.755	62	sin	0.883	75	sin	0.966
	cos	0.656		cos	0.469		cos	0.259
	tan	1.15		tan	1.88		tan	3.730
50	sin	0.766	63	sin	0.891	76	sin	0.970
	cos	0.643		cos	0.454		cos	0.242
	tan	1.19		tan	1.96		tan	4.010
51	sin	0.777	64	sin	0.899	77	sin	0.974
	cos	0.629		cos	0.438		cos	0.225
	tan	1.23		tan	2.05		tan	4.330
52	sin	0.788	65	sin	0.906	78	sin	0.978
	cos	0.616		cos	0.423		cos	0.208
	tan	1.28		tan	2.140		tan	4.700
53	sin	0.799	66	sin	0.914	79	sin	0.982
	cos	0.602		cos	0.407		cos	0.191
	tan	1.33		tan	2.250		tan	5.14
54	sin	0.809	67	sin	0.920	80	sin	0.985
	cos	0.588		cos	0.391		cos	0.174
	tan	1.38		tan	2.360		tan	5.67
55	sin	0.819	68	sin	0.927	81	sin	0.988
	cos	0.574		cos	0.375		cos	0.156
	tan	1.43		tan	2.480		tan	6.31
56	sin	0.829	69	sin	0.934	82	sin	0.990
	cos	0.559		cos	0.358		cos	0.139
	tan	1.48		tan	2.160		tan	7.12
57	sin	0.839	70	sin	0.940	83	sin	0.993
	cos	0.545		cos	0.342		cos	0.122
	tan	1.54		tan	2.750		tan	8.14
58	sin	0.848	71	sin	0.946	84	sin	0.995
	cos	0.530		cos	0.326		cos	0.105
	tan	1.60		tan	2.900		tan	9.51
59	sin	0.857	72	sin	0.951	85	sin	0.996
	cos	0.515		cos	0.309		cos	0.087
	tan	1.66		tan	3.080		tan	11.40
60	sin	0.866	73	sin	0.956	86	sin	0.998
	cos	0.500		cos	0.292		cos	0.070
	tan	1.73		tan	3.270		tan	14.30

Table 9-1. Trigonometric Functions (contd.)

87	sin	0.999	88	sin	0.999	89	sin	1.000	90	sin	1.000
	cos	0.052		cos	0.035		cos	0.017		cos	0.000
	tan	19.1		tan	28.6		tan	57.3		tan	

Table 9-1. Trigonometric Functions (contd.)

Example:

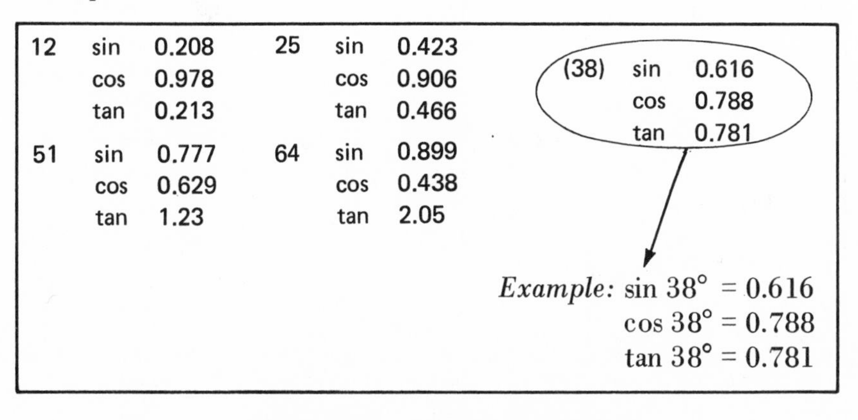

SECTION 2: LOGARITHMS

Tables for both Base 10 and Base e are provided. (See Table 9-2 and 9-3.) They are related by the following conversion equations:

$$\log_{10} X = \frac{\log_e X}{\log_e 10}$$

$$\log_e X = \frac{\log_{10} X}{\log_{10} e}$$

Table 9-2. Log Base 10

N	0	1	2	3	4	5	6	7	8	9
10	000	004	009	013	017	021	025	029	033	037
11	041	045	049	053	057	061	064	068	072	076
12	079	083	086	090	093	097	100	104	107	111
13	114	117	121	124	127	130	134	137	140	143
14	146	149	152	155	158	161	164	167	170	173
15	176	179	182	185	188	190	193	196	199	201
16	204	207	210	212	215	217	220	223	225	228
17	230	233	236	238	241	243	246	248	250	253
18	255	258	260	262	265	267	270	272	274	276
19	279	281	283	286	288	290	292	294	297	299

Table 9-2. Log Base 10 (contd.)

N	0	1	2	3	4	5	6	7	8	9
20	301	303	305	308	310	312	314	316	318	320
21	322	324	326	328	330	332	334	336	338	340
22	342	344	346	348	350	352	354	356	338	360
23	362	364	365	367	369	371	373	375	358	378
24	380	382	384	386	387	389	391	393	394	396
25	398	400	401	403	405	407	408	410	412	413
26	415	417	418	420	422	423	425	427	428	430
27	431	433	435	436	438	439	441	442	444	446
28	447	449	450	452	453	455	456	458	459	461
29	462	464	465	467	468	470	471	473	474	476
30	477	479	480	481	483	484	486	487	489	490
31	491	493	494	496	497	498	500	501	502	504
32	505	507	508	509	511	512	513	515	516	517
33	519	520	521	522	524	525	526	528	529	530
34	531	533	534	535	537	538	539	540	542	543
35	544	545	547	548	549	550	551	553	553	555
36	556	558	559	560	561	562	563	565	566	567
37	568	569	571	572	573	574	575	576	577	579
38	580	581	582	583	584	585	587	588	589	590
39	591	592	593	594	596	597	598	599	600	601
40	602	603	604	605	606	607	609	610	611	612
41	613	614	615	616	617	618	619	620	621	622
42	623	624	625	626	627	628	629	630	631	632
43	633	634	635	636	637	638	639	640	641	642
44	643	644	645	646	647	648	649	650	651	652
45	653	654	655	656	657	658	659	660	661	662
46	663	664	665	666	667	667	668	669	670	671
47	672	673	674	675	676	677	678	679	679	680
48	681	682	683	684	685	686	687	688	688	689
49	690	691	692	693	694	695	695	696	697	698
50	699	700	701	702	702	703	704	705	706	707
51	708	708	709	710	711	712	713	713	714	715
52	716	717	718	718	719	720	721	722	723	723
53	724	725	726	727	728	728	729	730	731	732
54	732	733	734	735	736	736	737	738	739	740
55	740	741	742	743	744	744	745	746	747	747

Table 9-2. Log Base 10 (contd.)

N	0	1	2	3	4	5	6	7	8	9
56	748	749	750	751	751	752	753	754	754	755
57	756	757	757	758	759	760	760	761	762	763
58	763	764	765	766	766	767	768	769	769	770
59	771	772	772	773	774	775	775	776	777	777
60	778	779	780	780	781	782	782	783	784	785
61	785	786	787	787	788	789	790	790	791	792
62	792	793	794	794	795	796	797	797	798	799
63	799	800	801	801	802	803	803	804	805	806
64	806	807	808	808	809	810	810	811	812	812
65	813	814	814	815	816	816	817	818	818	819
66	820	820	821	822	822	823	823	824	825	825
67	826	827	827	828	829	829	830	831	831	832
68	833	833	834	834	835	836	836	837	838	838
69	839	839	840	841	841	842	843	843	844	844
70	845	846	846	847	848	848	849	849	850	851
71	851	852	852	853	854	854	855	856	856	857
72	857	858	859	860	860	861	862	862	863	
73	863	864	865	865	866	866	867	867	868	869
74	869	870	870	871	872	872	873	873	874	874
75	875	876	876	877	877	878	879	879	880	880
76	881	881	882	883	883	884	884	885	885	886
77	886	887	888	888	889	889	890	890	891	892
78	892	893	893	894	894	895	895	896	897	897
79	898	898	899	899	900	900	901	901	902	903
80	903	904	904	905	905	906	906	907	907	908
81	908	909	910	910	911	911	912	912	913	913
82	914	914	915	915	916	916	917	918	918	919
83	919	920	920	921	921	922	922	923	923	924
84	924	925	925	926	926	927	927	928	928	929
85	929	930	930	931	931	932	932	933	933	934
86	934	935	936	936	937	937	938	938	939	939
87	940	940	941	941	942	942	942	943	943	944
88	944	945	945	946	946	947	947	948	948	949
89	949	950	950	951	951	952	952	953	953	954
90	954	955	955	956	956	957	957	958	958	959
91	959	960	960	960	961	961	962	962	963	963

Table 9-2. Log Base 10 (contd.)

N	0	1	2	3	4	5	6	7	8	9
92	964	964	965	965	966	966	967	967	968	968
93	968	969	969	970	970	971	971	972	972	973
94	973	974	974	975	975	975	976	976	977	977
95	978	978	979	979	980	980	980	981	981	982
96	982	983	983	984	984	985	985	985	986	986
97	987	987	988	988	989	989	989	990	990	991
98	991	992	992	993	993	993	994	994	995	995
99	996	996	997	997	997	998	998	999	999	000

Example:

N	0	1	2	3	4	5	6	7	8	9
20	301	303	305	308	310	312	314	316	318	320
21	322	324	326	328	330	332	334	336	338	340
22	342	344	346	348	350	352	354	356	358	360
23	362	364	365	367	369	371	373	375	377	378
24	380	382	384	386	387	389	391	393	394	396

Example:

$\log_{10} 22.7 = 1.356$

N	0	1	2	3	4	5	6	7	8	9
65	813	814	814	815	816	816	817	818	818	819
66	820	820	821	822	822	823	823	824	825	825
67	826	827	827	828	829	829	830	831	831	832
68	833	833	834	834	835	836	836	837	838	838
69	839	839	840	841	841	842	843	843	844	844

Example:

$\log_{10} 65.1 = 1.814$

Table 9-3. Log Base e

N	0	1	2	3	4	5	6	7	8	9
10	.000	.010	.020	.030	.039	.049	.058	.068	.077	.086
11	.095	.104	.113	.122	.131	.140	.148	.157	.166	.174
12	.182	.191	.199	.207	.215	.223	.231	.239	.247	.255
13	.262	.270	.278	.285	.293	.300	.307	.315	.322	.329
14	.336	.344	.351	.355	.365	.372	.378	.385	.392	.399
15	.405	.412	.419	.425	.432	.438	.445	.451	.457	.464
16	.470	.476	.482	.489	.495	.501	.507	.513	.519	.525
17	.531	.536	.542	.548	.554	.560	.565	.571	.577	.582
18	.558	.593	.599	.604	.610	.615	.621	.626	.631	.637
19	.642	.647	.652	.658	.663	.668	.673	.678	.683	.688
20	.693	.698	.703	.708	.713	.718	.723	.728	.732	.737
21	.742	.747	.751	.756	.761	.765	.770	.775	.779	.784
22	.788	.793	.798	.802	.806	.811	.815	.820	.824	.829
23	.833	.837	.842	.846	.850	.854	.859	.863	.867	.871
24	.875	.880	.884	.888	.892	.896	.900	.904	.908	.912
25	0.92	0.92	0.92	0.93	0.93	0.94	0.94	0.94	0.95	0.95
26	0.96	0.96	0.96	0.97	0.97	0.97	0.98	0.98	0.99	0.99
27	0.99	1.00	1.00	1.00	1.01	1.01	1.02	1.02	1.02	1.03
28	1.03	1.03	1.04	1.04	1.04	1.05	1.05	1.05	1.06	1.06
29	1.06	1.07	1.07	1.08	1.08	1.08	1.09	1.09	1.09	1.10
30	1.10	1.10	1.11	1.11	1.11	1.12	1.12	1.12	1.12	1.13
31	1.13	1.13	1.14	1.14	1.14	1.15	1.15	1.15	1.16	1.16
32	1.16	1.17	1.17	1.17	1.18	1.18	1.18	1.18	1.19	1.19
33	1.19	1.20	1.20	1.20	1.21	1.21	1.21	1.21	1.22	1.22
34	1.22	1.23	1.23	1.23	1.24	1.24	1.24	1.24	1.25	1.25
35	1.25	1.26	1.26	1.26	1.26	1.27	1.27	1.27	1.28	1.28
36	1.28	1.28	1.29	1.29	1.29	1.29	1.30	1.30	1.30	1.31
37	1.31	1.31	1.31	1.32	1.32	1.32	1.32	1.33	1.33	1.33
38	1.34	1.34	1.34	1.34	1.35	1.35	1.35	1.35	1.36	1.36
39	1.36	1.36	1.37	1.37	1.37	1.37	1.38	1.38	1.38	1.38
40	1.39	1.39	1.39	1.39	1.40	1.40	1.40	1.40	1.41	1.41
41	1.41	1.41	1.42	1.42	1.42	1.42	1.43	1.43	1.43	1.43
42	1.44	1.44	1.44	1.44	1.44	1.45	1.45	1.45	1.45	1.46
43	1.46	1.46	1.46	1.47	1.47	1.47	1.47	1.47	1.48	1.48
44	1.48	1.48	1.49	1.49	1.49	1.49	1.50	1.50	1.50	1.50

Table 9-3. Log Base e (contd.)

N	0	1	2	3	4	5	6	7	8	9
45	1.50	1.51	1.51	1.51	1.51	1.52	1.52	1.52	1.52	1.52
46	1.53	1.53	1.53	1.53	1.53	1.54	1.54	1.54	1.54	1.55
47	1.55	1.55	1.55	1.55	1.56	1.56	1.56	1.56	1.56	1.57
48	1.57	1.57	1.57	1.57	1.58	1.58	1.58	1.58	1.59	1.59
49	1.59	1.59	1.59	1.60	1.60	1.60	1.60	1.60	1.61	1.61
50	1.61	1.61	1.61	1.62	1.62	1.62	1.62	1.62	1.63	1.63
51	1.63	1.63	1.63	1.64	1.64	1.64	1.64	1.64	1.64	1.65
52	1.65	1.65	1.65	1.65	1.66	1.66	1.66	1.66	1.66	1.67
53	1.67	1.67	1.67	1.67	1.68	1.68	1.68	1.68	1.68	1.68
54	1.69	1.69	1.69	1.69	1.69	1.70	1.70	1.70	1.70	1.70
55	1.70	1.71	1.71	1.71	1.71	1.71	1.72	1.72	1.72	1.72
56	1.72	1.72	1.73	1.73	1.73	1.73	1.73	1.74	1.74	1.74
57	1.74	1.74	1.74	1.75	1.75	1.75	1.75	1.75	1.75	1.76
58	1.76	1.76	1.76	1.76	1.76	1.77	1.77	1.77	1.77	1.77
59	1.77	1.78	1.78	1.78	1.78	1.78	1.79	1.79	1.79	1.79
60	1.79	1.79	1.80	1.80	1.80	1.80	1.80	1.80	1.81	1.81
61	1.81	1.81	1.81	1.81	1.81	1.82	1.82	1.82	1.82	1.82
62	1.82	1.83	1.83	1.83	1.83	1.83	1.83	1.84	1.84	1.84
63	1.84	1.84	1.84	1.85	1.85	1.85	1.85	1.85	1.85	1.85
64	1.86	1.86	1.86	1.86	1.86	1.86	1.87	1.87	1.87	1.87
65	1.87	1.87	1.87	1.88	1.88	1.88	1.88	1.88	1.88	1.89
66	1.89	1.89	1.89	1.89	1.89	1.89	1.90	1.90	1.90	1.90
67	1.90	1.90	1.91	1.91	1.91	1.91	1.91	1.91	1.91	1.92
68	1.92	1.92	1.92	1.92	1.92	1.92	1.93	1.93	1.93	1.93
69	1.93	1.93	1.93	1.94	1.94	1.94	1.94	1.94	1.94	1.94
70	1.95	1.95	1.95	1.95	1.95	1.95	1.95	1.96	1.96	1.96
71	1.96	1.96	1.96	1.96	1.97	1.97	1.97	1.97	1.97	1.97
72	1.97	1.98	1.98	1.98	1.98	1.98	1.98	1.98	1.99	1.99
73	1.99	1.99	1.99	1.99	1.99	1.99	2.00	2.00	2.00	2.00
74	2.00	2.00	2.00	2.01	2.01	2.01	2.01	2.01	2.01	2.01
75	2.01	2.02	2.02	2.02	2.02	2.02	2.02	2.02	2.03	2.03
76	2.03	2.03	2.03	2.03	2.03	2.03	2.04	2.04	2.04	2.04
77	2.04	2.04	2.04	2.05	2.05	2.05	2.05	2.05	2.05	2.05
78	2.05	2.06	2.06	2.06	2.06	2.06	2.06	2.06	2.06	2.07
79	2.07	2.07	2.07	2.07	2.07	2.07	2.07	2.08	2.08	2.08
80	2.08	2.08	2.08	2.08	2.08	2.09	2.09	2.09	2.09	2.09
81	2.09	2.09	2.09	2.10	2.10	2.10	2.10	2.10	2.10	2.10
82	2.10	2.11	2.11	2.11	2.11	2.11	2.11	2.11	2.11	2.12

Table 9-3. Log Base e (contd.)

N	0	1	2	3	4	5	6	7	8	9
83	2.12	2.12	2.12	2.12	2.12	2.12	2.12	2.12	2.13	2.13
84	2.13	2.13	2.13	2.13	2.13	2.13	2.14	2.14	2.14	2.14
85	2.14	2.14	2.14	2.14	2.14	2.15	2.15	2.15	2.15	2.15
86	2.15	2.15	2.15	2.16	2.16	2.16	2.16	2.16	2.16	2.16
87	2.16	2.16	2.17	2.17	2.17	2.17	2.17	2.17	2.17	2.17
88	2.17	2.18	2.18	2.18	2.18	2.18	2.18	2.18	2.18	2.18
89	2.19	2.19	2.19	2.19	2.19	2.19	2.19	2.19	2.19	2.20
90	2.20	2.20	2.20	2.20	2.20	2.20	2.20	2.20	2.21	2.21
91	2.21	2.21	2.21	2.21	2.21	2.21	2.21	2.22	2.22	2.22
92	2.22	2.22	2.22	2.22	2.22	2.22	2.23	2.23	2.23	2.23
93	2.23	2.23	2.23	2.23	2.23	2.24	2.24	2.24	2.24	2.24
94	2.24	2.24	2.24	2.24	2.24	2.25	2.25	2.25	2.25	2.25
95	2.25	2.25	2.25	2.25	2.26	2.26	2.26	2.26	2.26	2.26
96	2.26	2.26	2.26	2.26	2.27	2.27	2.27	2.27	2.27	2.27
97	2.27	2.27	2.27	2.28	2.28	2.28	2.28	2.28	2.28	2.28
98	2.28	2.28	2.28	2.29	2.29	2.29	2.29	2.29	2.29	2.29
99	2.29	2.29	2.29	2.30	2.30	2.30	2.30	2.30	2.30	2.30

Example:

N	0	1	2	3	4	5	6	7	8	9
15	.405	.412	.419	.425	.432	.438	.445	.451	.457	.464
16	.470	.476	.482	.489	.495	.501	.507	.513	.519	.525
17	.531	.536	.542	.548	.554	.560	.565	.571	.577	.582
18	.588	.593	.599	.604	.610	.615	.621	.626	.631	.637
19	.642	.647	.652	.658	.663	.668	.673	.678	.683	.688

Example: $\log_e 1.90 = 0.642$ *Example:* $\log_e 1.55 = 0.438$

SECTION 3: POWERS OF TWO

Conversion of numbers from binary to decimal or from decimal to binary requires that the positive and negative powers of two be known. This table provides these values up to N = 20. (See Table 9-4.)

Table 9-4. Powers of 2

N	2^n	2^{-n}
0	1	1.0
1	2	0.50
2	4	0.25
3	8	0.125
4	16	0.0625
5	32	0.03125
6	64	0.015625
7	128	0.0078125
8	256	0.00390625
9	512	0.001953125
10	1024	0.0009765625
11	2048	0.00048828125
12	4096	0.000244140625
13	8192	0.0001220703125
14	16382	0.00006103515625
15	32768	0.000030517578125
16	65536	0.0000152587890625
17	131072	0.00000762939453125
18	262144	0.000003814697265625
19	524288	0.0000019073486328125
20	1048576	0.00000095367431640625

SECTION 4: DECIBELS

A decibel is literally 1/10 of a bel, a unit of measure named after Alexander Graham Bell. Decibels are defined by the following equations:

$$\text{Decibels (dB)} = 10 \text{ Log} \frac{P_1}{P_2}$$

$$\text{Decibels (dB)} = 20 \text{ Log} \frac{V_1}{V_2}$$

The logarithmic measure of the ratio of two values came about because the human ear responds logarithmically. One application of decibel measurements therefore is volume units (V.U.), where zero V.U. represents .001 watts across 600 ohms. (See Table 9-5.)

Table 9-5. Decibels

dB	+dB power ratio	-dB power ratio	+dB voltage ratio	-dB voltage ratio
0	1.00	1.00	1.00	1.00
1	1.26	0.794	1.1225	0.891
3	2.00	0.50	1.414	0.707
6	4.00	0.250	2.000	0.50
10	10.00	0.1	3.162	0.3162
20	100.00	0.01	10.00	0.10
30	1000.00	0.001	31.62	0.03162
40	10^{+4}	10^{-4}	10^{+2}	0.01
50	10^{+5}	10^{-5}	$3.162 \times 10^{+2}$	0.003162
60	10^{+6}	10^{-6}	10^{+3}	0.001
70	10^{+7}	10^{-7}	$3.162 \times 10^{+3}$	0.0003162
80	10^{+8}	10^{-8}	10^{+4}	0.0001

Example 1: A certain transmitter was operating with an output power of 10 watts. Its second harmonic output was down -43dB. What was the power content of the second harmonic?

Solution: A -43dB reduction is A -40dB reduction plus A -3dB reduction
A -40dB from 10 watts is $10 \times 10^{-4} = 10^{-3} = .001$ watts
Another -3dB is .001 watts $\times$ 1/2 = .0005 watts

Example 2: If 0dB is 600 milliwatts, how much is +26dB?

Solution: +26dB is 20dB + 6dB
600mw $\times$ 100 (20dB power) = 60 watts
60 watts $\times$ 4 (6dB power) = 240 watts

SECTION 5: METRIC CONVERSION

The metric system has an advantage over the English system since its units are all related by powers of ten. To change units, all that is necessary is to shift the decimal point. The following table gives the metric prefixes, along with their decimal multipliers. Also given are conversion factors between metric and English units for length, mass, and volume. (See Table 9-6.)

Table 9-6. Metric System

Multiples		Prefixes
1,000,000,000,000	-10^{12}	TERA
1,000,000,000	-10^{9}	GIGA
1,000,000	-10^{6}	MEGA
1,000	-10^{3}	KILO
100	-10^{2}	HECTO
10	-10^{1}	DEKA
0.1	-10^{-1}	DECI
0.01	-10^{-2}	CENTI
0.001	-10^{-3}	MILLI
0.000001	-10^{-6}	MICRO
0.000000001	-10^{-9}	NANO
0.000000000001	-10^{-12}	PICO

LENGTH

Centimeter = 0.3937 inch	Inch = 2.54 centimeter
Meter = 3.28 feet	Foot = 0.3048 meter
Kilometer = 0.621 mile	Mile = 1.61 kilometer

MASS

Gram = 0.0353 ounce	Ounce = 28.35 grams
Kilogram = 2.205 lb	Lb = 0.436 kilograms

VOLUME

Liter = 1.057 quarts	Quart = 0.946 liter
Meter$^{(3)}$ = 35.3 Ft$^{(3)}$	Ft$^{(3)}$ = 0.0283 meter$^{(3)}$

SECTION 6: HEX TO DECIMAL CONVERSION

Methods used to convert between base systems are given in Chapter 1. Conversion between hex and decimal is very common, and the following table can be quicker and easier than the methods shown in Chapter 1.

N	0	1	2	3	4	5	6	7	8	9	A	B	C	D	E	F
0	0000	0001	0002	0003	0004	0005	0006	0007	0008	0009	0010	0011	0012	0013	0014	0015
1	0016	0017	0018	0019	0020	0021	0022	0023	0024	0025	0026	0027	0028	0029	0030	0031
2	0032	0033	0034	0035	0036	0037	0038	0039	0040	0041	0042	0043	0044	0045	0046	0047
3	0048	0049	0050	0051	0052	0053	0054	0055	0056	0057	0058	0059	0060	0061	0062	0063
4	0064	0065	0066	0067	0068	0069	0070	0071	0072	0073	0074	0075	0076	0077	0078	0079
5	0080	0081	0082	0083	0084	0085	0086	0087	0088	0089	0090	0091	0092	0093	0094	0095
6	0096	0097	0098	0099	0100	0101	0102	0103	0104	0105	0106	0107	0108	0109	0110	0111
7	0112	0113	0114	0115	0116	0117	0118	0119	0120	0121	0122	0123	0124	0125	0126	0127
8	0128	0129	0130	0131	0132	0133	0134	0135	0136	0137	0138	0139	0140	0141	0142	0143
9	0144	0145	0146	0147	0148	0149	0150	0151	0152	0153	0154	0155	0156	0157	0158	0159
A	0160	0161	0162	0163	0164	0165	0166	0167	0168	0169	0170	0171	0172	0173	0174	0175
B	0176	0177	0178	0179	0180	0181	0182	0183	0184	0185	0186	0187	0188	0189	0190	0191
C	0192	0193	0194	0195	0196	0197	0198	0199	0200	0201	0202	0203	0204	0205	0206	0207
D	0208	0209	0210	0211	0212	0213	0214	0215	0216	0217	0218	0219	0220	0221	0222	0223
E	0224	0225	0226	0227	0228	0229	0230	0231	0232	0233	0234	0235	0236	0237	0238	0239
F	0240	0241	0242	0243	0244	0245	0246	0247	0248	0249	0250	0251	0252	0253	0254	0255
10	0256	0257	0258	0259	0260	0261	0262	0263	0264	0265	0266	0267	0268	0269	0270	0271
11	0272	0273	0274	0275	0276	0277	0278	0279	0280	0281	0282	0283	0284	0285	0286	0287
12	0288	0289	0290	0291	0292	0293	0294	0295	0296	0297	0298	0299	0300	0301	0302	0303
13	0304	0305	0306	0307	0308	0309	0310	0311	0312	0313	0314	0315	0316	0317	0318	0319
14	0320	0321	0322	0323	0324	0325	0326	0327	0328	0329	0330	0331	0332	0333	0334	0335
15	0336	0337	0338	0339	0340	0341	0342	0343	0344	0345	0346	0347	0348	0349	0350	0351
16	0352	0353	0354	0355	0356	0357	0358	0359	0360	0361	0362	0363	0364	0365	0366	0367
17	0368	0369	0370	0371	0372	0373	0374	0375	0376	0377	0378	0738	0380	0381	0382	0383
18	0384	0385	0386	0387	0388	0389	0390	0391	0392	0393	0394	0395	0396	0397	0398	0399

Table 9-7. Hex to Decimal Conversion Table

N	0	1	2	3	4	5	6	7	8	9	A	B	C	D	E	F
19	0400	0401	0402	0403	0404	0405	0406	0407	0408	0409	0410	0411	0412	0413	0414	0415
1A	0416	0417	0418	0419	0420	0421	0422	0423	0424	0425	0426	0427	0428	0429	0430	0431
1B	0432	0433	0434	0435	0436	0437	0438	0439	0440	0441	0442	0443	0444	0445	0446	0447
1C	0448	0449	0450	0451	0452	0453	0454	0455	0456	0457	0458	0459	0460	0461	0462	0463
1D	0464	0465	0466	0467	0468	0469	0470	0471	0472	0473	0474	0475	0476	0477	0478	0479
1E	0480	0481	0482	0483	0484	0485	0486	0487	0488	0489	0490	0491	0492	0493	0494	0495
1F	0496	0497	0498	0499	0500	0501	0502	0503	0504	0505	0506	0507	0508	0509	0510	0511
20	0512	0513	0514	0515	0516	0517	0518	0519	0520	0521	0522	0523	0524	0525	0526	0527
21	0528	0529	0530	0531	0532	0533	0534	0535	0536	0537	0538	0539	0540	0541	0542	0543
22	0544	0545	0546	0547	0548	0549	0550	0551	0552	0553	0554	0555	0556	0557	0558	0559
23	0560	0561	0562	0563	0564	0565	0566	0567	0568	0569	0570	0571	0572	0573	0574	0575
24	0576	0577	0578	0579	0580	0581	0582	0583	0584	0585	0586	0587	0588	0589	0590	0591
25	0592	0593	0594	0595	0596	0597	0598	0599	0600	0601	0602	0603	0604	0605	0606	0607
26	0608	0609	0610	0611	0612	0613	0614	0615	0616	0617	0618	0619	0620	0621	0622	0623
27	0624	0625	0626	0627	0628	0629	0630	0631	0632	0633	0634	0635	0636	0637	0638	0639
28	0640	0641	0642	0643	0644	0645	0646	0647	0648	0649	0650	0651	0652	0653	0654	0655
29	0656	0657	0658	0659	0660	0661	0662	0663	0664	0665	0666	0667	0668	0669	0670	0671
2A	0672	0673	0674	0675	0676	0677	0678	0679	0680	0681	0682	0683	0684	0685	0686	0687
2B	0688	0689	0690	0691	0692	0693	0694	0695	0696	0697	0698	0699	0700	0701	0702	0703
2C	0704	0705	0706	0707	0708	0709	0710	0711	0712	0713	0714	0715	0716	0717	0718	0719
2D	0720	0721	0723	0723	0724	0725	0726	0727	0728	0729	0730	0731	0732	0733	0734	0735
2E	0736	0737	0738	0739	0740	0741	0742	0743	0744	0745	0746	0747	0748	0749	0750	0751
2F	0752	0753	0754	0755	0756	0757	0758	0759	0760	0761	0762	0763	0764	0765	0766	0767
30	0768	0769	0770	0771	0772	0773	0774	0775	0776	0777	0778	0779	0780	0781	0782	0783

Table 9-7. Hex to Decimal Conversion Table (contd.)

N	0	1	2	3	4	5	6	7	8	9	A	B	C	D	E	F
31	0784	0785	0786	0787	0788	0789	0790	0791	0792	0793	0794	0795	0796	0797	0798	0799
32	0800	0801	0802	0803	0804	0805	0806	0807	0808	0809	0810	0811	0812	0813	0814	0815
33	0816	0817	0818	0819	0820	0821	0822	0823	0824	0825	0826	0827	0828	0829	0830	0831
34	0832	0833	0834	0835	0836	0837	0838	0839	0840	0841	0842	0843	0844	0845	0846	0847
35	0848	0849	0850	0851	0852	0853	0854	0855	0856	0857	0858	0859	0860	0861	0862	0863
36	0864	0865	0866	0867	0868	0869	0870	0871	0872	0873	0874	0875	0876	0877	0878	0879
37	0880	0881	0882	0883	0884	0885	0886	0887	0888	0889	0890	0891	0892	0893	0894	0895
38	0896	0897	0898	0899	0900	0901	0902	0903	0904	0905	0906	0907	0908	0909	0910	0911
39	0912	0913	0914	0915	0916	0917	0918	0919	0920	0921	0922	0923	0924	0925	0926	0927
3A	0928	0929	0930	0931	0932	0933	0934	0935	0936	0937	0938	0939	0940	0941	0942	0943
3B	0944	0945	0946	0947	0948	0949	0950	0951	0952	0953	0954	0955	0956	0957	0958	0959
3C	0960	0961	0962	0963	0964	0965	0966	0967	0968	0969	0970	0971	0972	0973	0974	0975
3D	0976	0977	0978	0979	0980	0981	0982	0983	0984	0985	0986	0987	0988	0989	0990	0991
3E	0992	0993	0994	0995	0996	0997	0998	0999	1000	1001	1002	1003	1004	1005	1006	1007
3F	1008	1009	1010	1011	1012	1013	1014	1015	1016	1017	1018	1019	1020	1021	1022	1023

Table 9-7. Hex to Decimal Conversion Table (contd.)

Example:

N	0	1	2	3	4	5	6	7	8	9	A	B	C	D	E	F
14	0320	0321	0322	0323	0324	0325	0326	0327	0328	0329	0330	0331	0332	0333	0334	0335
15	0336	0337	0338	0339	0340	0341	0342	0343	0344	0345	0346	0347	0348	0349	0350	0351

156 hex = 342 decimal
142 hex = 322 decimal

SECTION 7: ASCII CODE

The ASCII code is probably the most frequently encountered code in digital electronics. The following chart allows quick reference for each even seven-bit number and its alphanumeric equivalent. (See Table 9-8.)

Table 9-8. ASCII Code

	000	001	010	011	100	101	110	111
0000	NULL	DC_0 ①	ƀ	0	@	P		
0001	SOM	DC_1	!	1	A	Q		
0010	EOA	DC_2	"	2	B	R		
0011	EOM	DC_3	#	3	C	S		
0100	EOT	DC_4 STOP	$	4	D	T		
0101	WRU	ERR	%	5	E	U		
0110	RU	SYNC	&	6	F	V		
0111	BELL	LEM	'	7	G	W		
1000	FE_0	S_0	(	8	H	X		
1001	HT / SK	S_1	)	9	I	Y		
1010	LF	S_2	*	:	J			
1011	V_{TAB}	S_3	+	;	K	[		
1100	FF	S_4	comma ,	<	L	\		ACK
1101	CR	S_5	—	=	M	]		2
1110	SO	S_6	★	>	N	↑		ESC
1111	SI	S_7	/	?	O	←		DEL

(Blank spaces are unassigned.)

	top	side		
Example:	100	1000	=	H

Table 9-8A. ASCII Code Abbreviations

NULL–Null idle	DC_1–DC_2–Device control
SOM–Start of message	ERR–Error
EOA–End of address	SYNC–Synchronous idle
EOM–End of message	LEM–Logical end of media
EOT–End of transmission	ACK–Acknowledge
WRU–Who are you?	–Unassigned control
RU–Are you?	ESC–Escape
BELL–Audible bell	DEL–Delete idle
FE–Format effector	SO_0-SO_7–Separators
HT–Horizontal tabulation	
SK–Skip	
LF–Line feed	
VTAB–Vertical tabulation	
DCO–Device control	
FF–Form feed	
CR–Carriage return	
SO–Shift out	
SI–Shift in	

SECTION 8: E.B.C.D.I.C. CODE

IBM series 360 and 370 computer systems, as well as many other large systems, use a code known as ebcdic. The following table gives the seven-bit number and the corresponding alphanumeric character for the ebcdic code. (See Table 9-9 on pages 220 and 221.)

SECTION 9: BINARY TO DECIMAL CONVERSION

Conversions between binary and decimal numbers can be done by the methods outlined in Chapter 1, or they can be done quickly with the use of the following table. (See Table 9-10 on pages 222 and 223.)

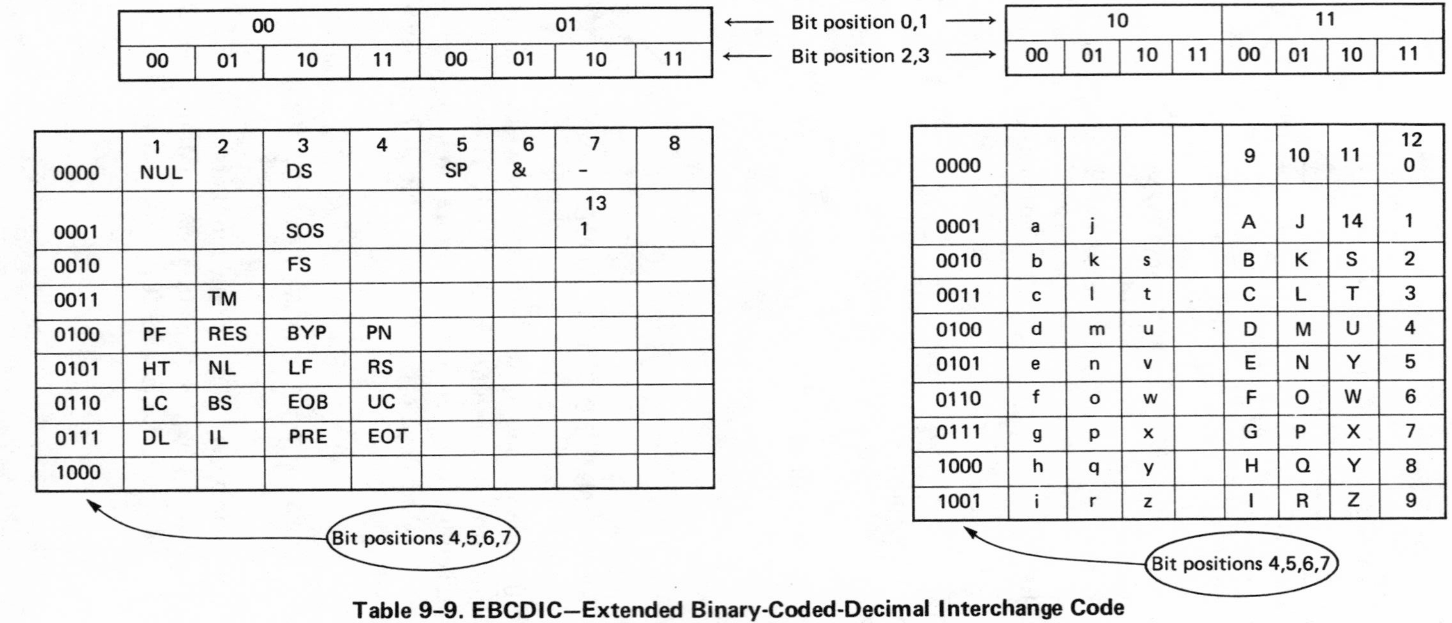

Bit positions 4,5,6,7	00				01			
	00	01	10	11	00	01	10	11
0000	1 NUL	2	3 DS	4	5 SP	6 &	7 -	8
0001			SOS				13 1	
0010			FS					
0011		TM						
0100	PF	RES	BYP	PN				
0101	HT	NL	LF	RS				
0110	LC	BS	EOB	UC				
0111	DL	IL	PRE	EOT				
1000								

Bit positions 4,5,6,7	10				11			
	00	01	10	11	00	01	10	11
0000					9	10	11	12 0
0001	a	j			A	J	14	1
0010	b	k	s		B	K	S	2
0011	c	l	t		C	L	T	3
0100	d	m	u		D	M	U	4
0101	e	n	v		E	N	Y	5
0110	f	o	w		F	O	W	6
0111	g	p	x		G	P	X	7
1000	h	q	y		H	Q	Y	8
1001	i	r	z		I	R	Z	9

Bit position 0,1 — Bit position 2,3

Table 9–9. EBCDIC—Extended Binary-Coded-Decimal Interchange Code

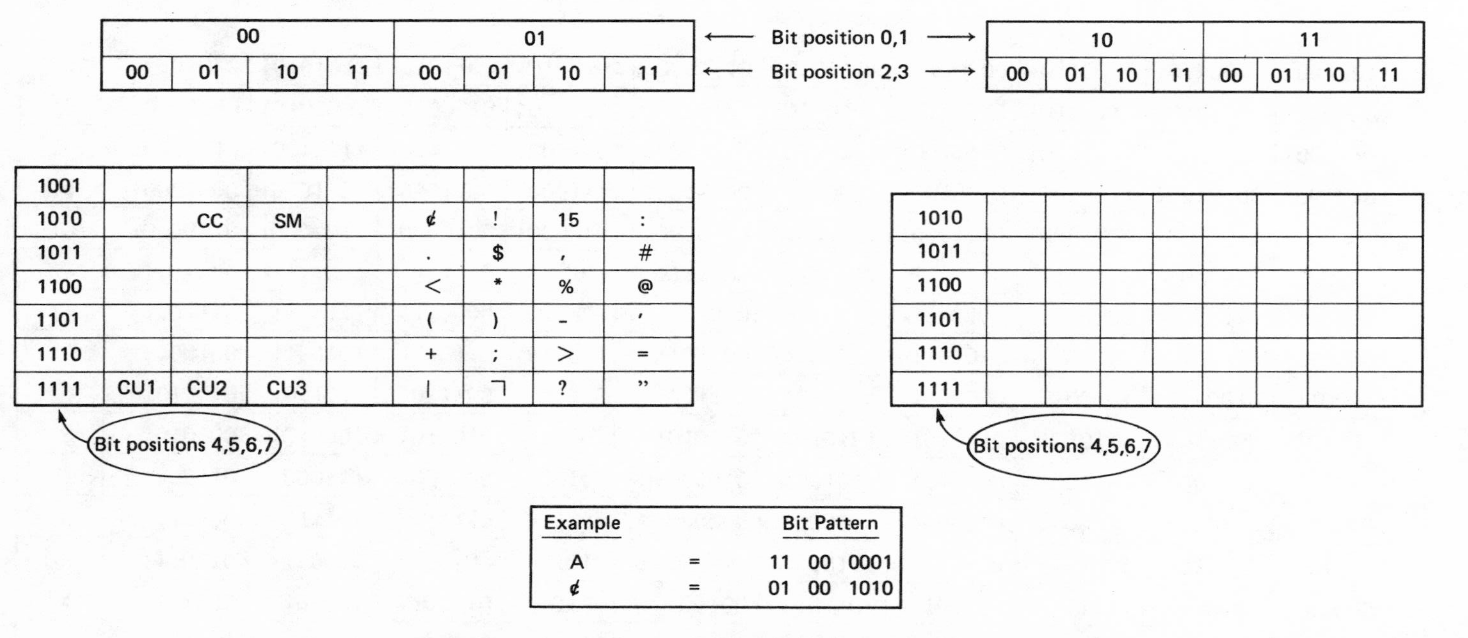

Bit position 0,1 →	00				01			
Bit position 2,3 →	00	01	10	11	00	01	10	11
1001								
1010		CC	SM		¢	!	15	:
1011					.	$	,	#
1100					<	*	%	@
1101					(	)	-	'
1110					+	;	>	=
1111	CU1	CU2	CU3		|	┐	?	"

Bit positions 4,5,6,7

Bit position 0,1 →	10				11			
Bit position 2,3 →	00	01	10	11	00	01	10	11
1010								
1011								
1100								
1101								
1110								
1111								

Bit positions 4,5,6,7

Example		Bit Pattern
A	=	11 00 0001
¢	=	01 00 1010

Table 9-9. EBCDIC—Extended Binary-Coded-Decimal Interchange Code (contd.)

	0	1	2	3	4	5	6	7	8	9
0	0	1	10	11	100	101	110	111	1000	1001
1	1010	1011	1100	1101	1110	1111	10000	10001	10010	10011
2	10100	10101	10110	10111	11000	11001	11010	11011	11100	11101
3	11110	11111	100000	100001	100010	100011	100100	100101	100110	100111
4	101000	101001	101010	101011	101100	101101	101110	101111	110000	110001
5	110010	110011	110100	110101	110110	110111	111000	111001	111010	111011
6	111100	111101	111110	111111	1000000	1000001	1000010	1000011	1000100	1000101
7	1000110	1000111	1001000	1001001	1001010	1001011	1001100	1001101	1001110	1001111
8	1010000	1010001	1010010	1010011	1010100	1010101	1010110	1010111	11011000	1011001
9	1011010	1011011	1011100	1011101	1011110	1011111	1100000	1100001	1100010	1100011
10	1100100	1100101	1100110	1100111	1101000	1101001	1101010	1101011	1101100	1101101
11	1101110	1101111	1110000	1110001	1110010	1110011	1110100	1110101	1110110	1110111
12	1111000	1111001	1111010	1111011	1111100	1111101	1111110	1111111	10000000	10000001
13	10000010	10000011	10000100	10000101	10000110	10000111	10001000	10001001	10001010	10001011
14	10001100	10001101	10001110	10001111	10010000	10010001	10010010	10010011	10010100	10010101
15	10010110	10010111	10011000	10011001	10011010	10011011	10011100	10011101	10011110	10011111
16	10100000	10100001	10100010	10100011	101001000	10100101	10100110	10100111	10101000	10101001
17	10101010	10101011	10101100	10100101	10101110	10101111	10110000	10110001	10110010	10110011

Table 9-10. Binary to Decimal Conversion Table

	0	1	2	3	4	5	6	7	8	9
18	10110100	10110101	10110110	10110111	10111000	10111001	10111010	10111011	10111100	10111101
19	0111110	10111111	11000000	11000001	11000010	11000011	11000100	11000101	11000110	11000111
20	11001000	11001001	11001010	11001011	11001100	11001101	11001110	11001111	11010000	11010001
21	11010010	11010011	11010100	11010101	11010110	11010111	11011000	11011001	11011010	11011011
22	11011100	11011101	11011110	11011111	11100000	11100001	11100010	11100011	11100100	11100101
23	11100110	11100111	11101000	11101001	11101010	11101011	11101100	11101101	11101110	11101111
24	11110000	11110001	11110010	11110011	11110100	11110101	11110110	11110111	11111000	11111001
25	11111010	11111011	11111100	11111101	11111110	11111111				

Table 9-10. Binary to Decimal Conversion Table (contd.)

Example:

	0	1	2	3	4	5	6	7	8	9
13	10000010	10000011	10000100	10000101	10000110	10000111	10001000	10001001	10001010	10001011
14	0001100	10001101	10001110	10001111	10010000	10010001	10010010	10010011	10010100	10010101

Example: $146_{10} = 10010010_2$

Example: $132_{10} = 10000100_2$

SECTION 10: CONSTANTS AND SYMBOLS

The following is a list of constants and math symbols that are frequently encountered in electronics.

Table 9-11

CONSTANTS	SYMBOLS
$\pi = 3.14159$	$=$ equal
$2\pi = 6.28319$	$\cong$ approximately equal
$(2\pi)^2 = 39.47842$	$\neq$ not equal
$4\pi = 12.56637$	$\equiv$ identity
$\pi^2 = 9.86960$	$\perp$ perpendicular to
$\sqrt{\pi} = 1.77245$	$\parallel$ parallel to
$\frac{1}{\sqrt{\pi}} = 0.56419$	$\lvert x \rvert$ absolute value of X
$\frac{\pi}{2} = 1.57080$	$>$ is greater than
$\frac{1}{\pi} = 0.31831$	$\gg$ is much greater than
$\frac{1}{2\pi} = 0.15915$	$<$ is less than
$\frac{1}{\pi^2} = 0.10132$	$\ll$ is much less than
$\frac{\pi}{2} = 1.25331$	$\geq$ greater than or equal
$\sqrt{2} = 1.41421$	$\leq$ less than or equal
$\frac{1}{\sqrt{2}} = 0.70711$	$+$ add
$\text{Log}_e \pi = 1.14473$	$-$ subtract
$\text{Log}_{10} \pi = 0.49715$	$\times$ multiply
	$\div$ divide
	$\pm$ positive or negative

Euler's constant	$= 0.57722$
electron charge	$= 1.602 \times 10^{-19}$
1 coulomb	$= 6.25 \times 10^{18}$ electrons
1 ampere	= 1 coulomb/sec
1 farad	= 1 coulomb/volt
1 henry	= 1 voltsecond/ampere

This chapter contains practical information on several special problems found in electronic math, with sections written as reference guides for people involved in electronics. There are solutions given for "mind-teaser" type problems such as a cube made from 1-OHM resistors, as well as solutions for designing a transistor amplifier to meet a predetermined gain.

There is a section that derives the formula for gain in negative feedback amplifiers, including op-amps. There is also a section explaining how I and Q signals are used to transmit and reproduce the proper color information in T.V. broadcasting. The section on the single sideband shows, step by step, how the unwanted sideband is eliminated by proper phasing of the input signals. Each section is written as a quick reference source for the formula and methods associated with each subject:

SECTION 1: THE ONE-OHM CUBE

One of the most common problems found on electronic math quizzes and mind-teaser exams involves a cube having one-ohm resistors in each leg of the cube. (See Figure 10-1.)

Example:

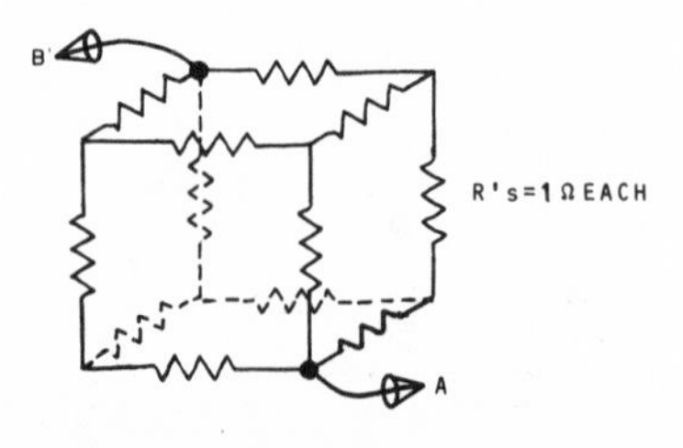

Figure 10-1

What is the total resistance between points A and B?

There are two methods of solution, both based on the symmetry of the cube.

Solution 1 (See Figure 10-2.)

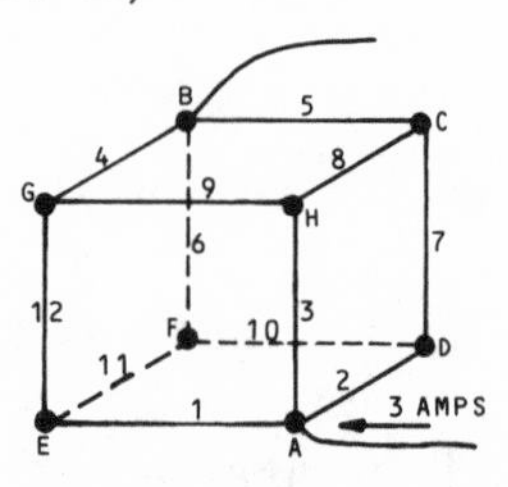

Figure 10-2

Assume three amps are flowing into the cube at A and out at B. Since the cube is symmetrical, the current splits evenly into Resistors 1, 2, and 3. Resistors 1, 2, and 3 each carry one amp, and the voltage drop across each is one volt. By the same token, Resistors 4, 5, and 6 each carry one amp, and the voltage drop across each is one volt. The vertices E, D, and H are at the same potential, and G, C, and F are at the same potential.

Considering point D, the amp entering this point splits equally, a half-amp through R10 and a half-amp through R7. The voltage drop across either resistor is a half-volt. The total voltage drop across the cube is:

$$1 \text{ volt} + 1 \text{ volt} + \frac{1}{2} \text{ volt} = 2.5 \text{ volts}$$

$$\text{Total R} = \frac{2.5}{3} = .8333 \text{ ohms}$$

Solution 2 (See Figure 10-3.)

When two or more points in a circuit are at equal potential, the points can be tied together with no change in the circuit. Points E, D, and H are at the same potential, as are G, C, and F. The circuit can be redrawn as below:

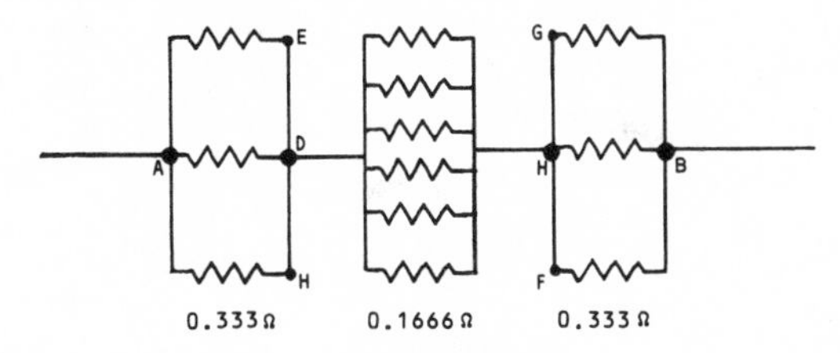

Figure 10-3

$$R_t = 0.333 + 0.333 + .1666 = .8333\Omega$$

SECTION 2: A RESISTOR LADDER NETWORK

Another one of the most common type of problems on electronic math quizzes and other mind-teasers is an infinite resistor ladder network. (See Figure 10-4.)

Example:

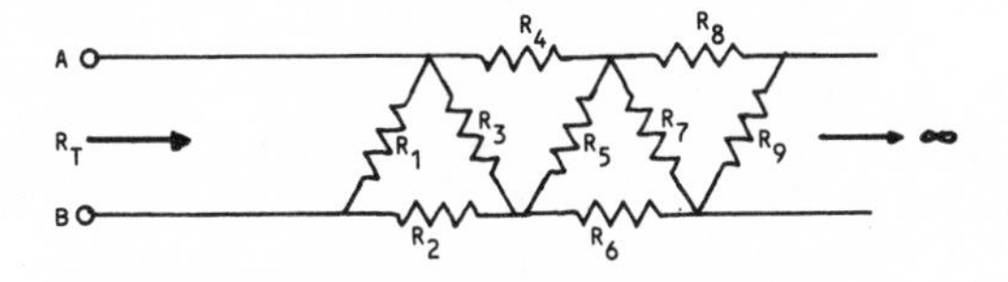

Figure 10-4

$$R_1 \ \& \ R_2 \ \& \ \ldots . = 1\Omega$$

What is the total resistance between Points A and B? (A) It is obvious that $R_t < 1$, since there is a 1-ohm resistor in parallel with the rest of the ladder. (B) Close approximations can be made by terminating the ladder and calculating the resistance.

First approximation: (See Figure 10-5.)

$$R_T = \frac{2}{1+2} = \frac{2}{3} = 0.666\Omega$$

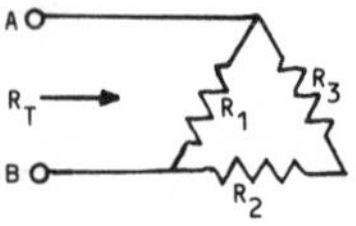

Figure 10-5

Second approximation: (See Figure 10-6.)

$$R_4 + R_5 = 2$$

$$2 \parallel R_3 = \frac{2}{1+2} = 0.666$$

$$0.666 + R_2 = 1.666$$

$$1.666 \parallel R_1 = \frac{1.666}{2.666} = 0.6249\Omega$$

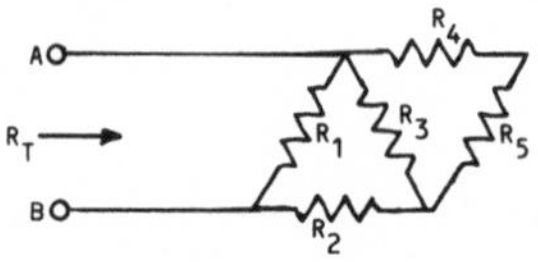

Figure 10-6

These approximations are getting close to the correct answer, which must be slightly below 0.6249. The exact value can be found as shown in Figure 10-7.

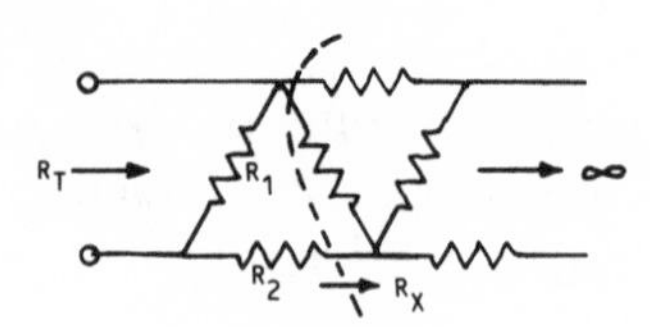

Figure 10-7

Divide the ladder as shown on the dotted line and call the remaining resistance R_x. The value of R_t now becomes R_1 in parallel with $R_2 + R_x$.

$$R_t = \frac{R_1(R_2 + R_x)}{R_1 + R_2 + R_x}$$

Since the ladder goes to infinity, the values R_x and R_t must be equal.

$$R_t = \frac{R_1(R_2 + R_x)}{R_1 + R_2 + R_x} = R_x$$

$$R_x = \frac{R_1(R_2 + R_x)}{R_1 + R_2 + R_x}$$

$$\cancel{R_1 R_x} + R_x R_2 + (R_x)^2 = R_1 R_2 + \cancel{R_1 R_x}$$

$$(R_x)^2 + R_2 R_x - R_1 R_2 = 0$$

$$R_x = \frac{-R_2 \pm \sqrt{(R_2)^2 + 4R_1 R_2}}{2} \qquad \text{Quadratic equation}$$

Since all R's = 1:

$$R_x = \frac{-1 \pm \sqrt{1+5}}{2} = \frac{-1 \pm 2.23607}{2}$$

Assuming R cannot be negative:

$$R_x = \frac{-1 + 2.23607}{2} = 0.618035 \text{ ohms}$$

SECTION 3: S.S.B. PHASING SYSTEM

Single sideband signals can be generated by proper phasing of the A.F. and R.F. components prior to the modulator. When the component frequencies are divided into two signals 90 degrees apart and combined after the modulator, the unwanted mixing products can be eliminated.

Amplitude Modulation

The process of modulation is the same as multiplication of the two signals. (See Figure 10-8.)

Example:

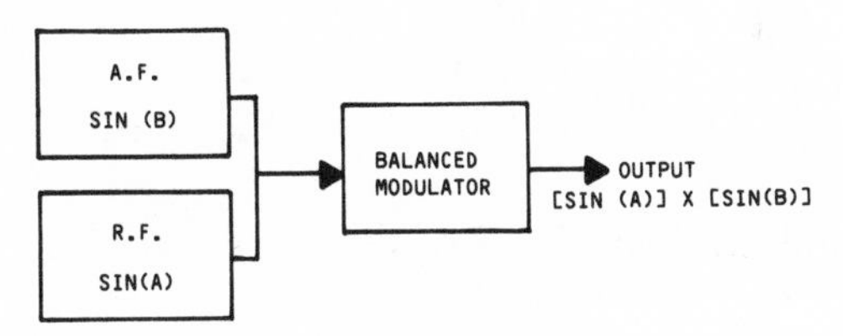

Figure 10-8

The balanced modulator cancels the R.F. (Sin A) signal in the output. The two signals being fed to the modulator are the audio Sin (B) and R.F. Sin (A), where (B) represents the frequency of the audio signal and (A) represents the R.F. frequency.

$$\text{Output} = \text{Sin (A)} \times \text{Sin (B)}$$

Trig Identities

$$2 \text{ Sin A Cos B} = \text{Sin } (A + B) + \text{Sin } (A - B)$$
$$2 \text{ Cos A Cos B} = \text{Cos } (A + B) + \text{Cos } (A - B)$$
$$2 \text{ Sin A Sin B} = \text{Cos } (A - B) - \text{Cos } (A + B)$$
$$2 \text{ Cos A Sin B} = \text{Sin } (A + B) - \text{Sin } (A - B)$$

Using these identities:

$$\text{Output} = \text{Sin A Sin B} = \frac{1}{2} [\text{Cos } (A - B) - \text{Cos } (A + B)]$$

Cos (A - B) and Cos (A + B) are the mixing products of the modulator. The frequencies of these products are:

(A - B) The R.F. frequency minus the A.F. frequency, or the lower sideband.

(A + B) The R.F. frequency plus the A.F. frequency, or the upper sideband.

This output is a double-sideband suppressed carrier signal. If the modulator had not been a balanced type, the R.F. signal Sin (A) would have appeared in the output.

$$\text{Output} = \underbrace{\text{Sin (A)}}_{\text{Carrier}} + \frac{1}{2} [\underbrace{\text{Cos } (A - B)}_{\text{Lower Sideband}} - \underbrace{\text{Cos } (A + B)}_{\text{Upper Sideband}}]$$

This is the standard A.M. signal of the carrier and upper/lower sidebands.

S.S.B. Phasing

A block diagram of S.S.B. phasing system is shown in Figure 10-9. This example produces a lower sideband signal. If an upper sideband signal is desired, it can be done by switching either the A.F. lines or the R.F. lines.

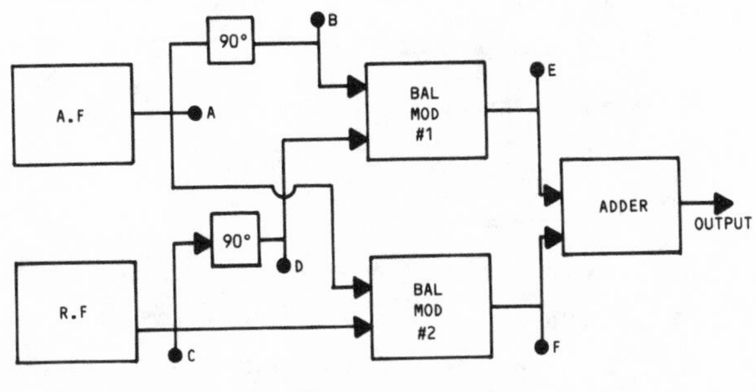

Figure 10-9

Let: $A = \text{Sin}\ (B)$
$C = \text{Sin}\ (A)$

Therefore: $B = \text{Cos}\ (B)$
$D = \text{Cos}\ (A)$

$$E = \text{Cos}\ (A)\ \text{Cos}\ (B) = \frac{1}{2}\ [\text{Cos}\ (A + B) + \text{Cos}\ (A - B)]$$

$$F = \text{Sin}\ (A)\ \text{Sin}\ (B) = \frac{1}{2}\ [\text{Cos}\ (A - B) - \text{Cos}\ (A + B)]$$

The upper sideband Cos (A + B) is positive at E and negative at F. When the two signals are combined, the upper sideband disappears.

$$\text{Output} = E + F = \text{Cos}\ (A - B)$$

The Third Method

The so-called "Third Method" of generating an S.S.B. signal also uses phasing to eliminate the unwanted mixing products. A block diagram is shown in Figure 10-10.

The audio frequency oscillator is set to a frequency in the

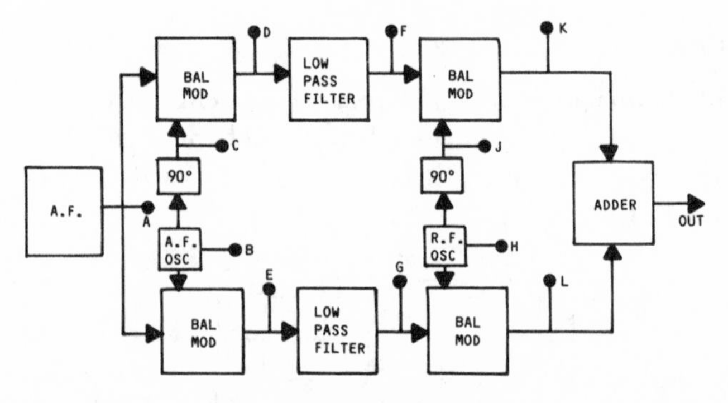

Figure 10-10

center of the audio bandpass, for example, 1500 HZ. This means that there are two cases: one where the audio signal at A is less than 1500 HZ, and two where the signal at A is greater than 1500 HZ.

Case 1: Audio frequency less than 1500 HZ.

Let: $A = \text{Sin}\ (B)$
$B = \text{Sin}\ (A)$

Therefore: $E = AB = \text{Sin}\ (B)\ \text{Sin}\ (A)$
$C = \text{Cos}\ (A)$
$D = \text{Sin}\ (B)\ \text{Cos}\ (A)$

Trig I.D.: $D = \frac{1}{2}\ [\text{Sin}\ (A + B) - \text{Sin}\ (A - B)]$

$$E = \frac{1}{2}\ [\text{Cos}\ (A - B) - \text{Cos}\ (A + B)]$$

The low-pass filters are used to eliminate the frequencies above 1500 HZ. This means that the A + B components are filtered out.

Therefore: $F = \frac{1}{2}\ [-\text{Sin}\ (A - B)]$

$$G = \frac{1}{2}\ [\text{Cos}\ (A - B)]$$

These two signals are 90 degrees apart and contain no frequencies above 1500 HZ. Since the half in front of the equations represents the amplitude of the signal, we will drop it to simplify the following:

$$F = -\mathrm{Sin}\,(A - B)$$
$$G = \mathrm{Cos}\,(A - B)$$

Let: $H = \mathrm{Sin}\,(C)$
$J = \mathrm{Cos}\,(C)$

Therefore: $K = FJ = \mathrm{Cos}\,(C)\,[-\mathrm{Sin}\,(A - B)]$
$L = GH = \mathrm{Sin}\,(C)\,[\mathrm{Cos}\,(A - B)]$

Trig I.D.: $L = \frac{1}{2}\,[\mathrm{Sin}\,(C + (A - B)) + \mathrm{Sin}\,(C - (A - B))]$

$$K = -\frac{1}{2}\,[\mathrm{Sin}\,(C + (A - B)) - \mathrm{Sin}\,(C - (A - B))]$$

$$K = \frac{1}{2}\,[-\mathrm{Sin}\,(C + (A - B)) + \mathrm{Sin}\,(C - (A - B))]$$

Output $= K + L = \mathrm{Sin}\,[C - (A - B)]$

This is a signal below the R.F. carrier C, except when the audio signal frequency equals the audio oscillator frequency $(A - B) = 0$. The frequency is then C.

Case 2: Audio frequency greater than 1500 HZ.

The signals at D and E were:

$$D = \frac{1}{2}\,[\mathrm{Sin}\,(A + B) - \mathrm{Sin}\,(A - B)]$$

$$E = \frac{1}{2}\,[\mathrm{Cos}\,(A - B) - \mathrm{Cos}\,(A + B)]$$

and at F and G:

$$F = \frac{1}{2}\,[-\mathrm{Sin}\,(A - B)]$$

$$G = \frac{1}{2}\,[\mathrm{Cos}\,(A - B)]$$

But now $B \geq A$, so $(A - B)$ becomes negative:

$$\sin(-x) = -\sin x$$
$$\cos(-x) = \cos x$$

So: $$F = \frac{1}{2}[\sin(A - B)]$$

$$G = \frac{1}{2}[\cos(A - B)]$$

Or: $$F = \sin(A - B)$$
$$G = \cos(A - B)$$
$$K = FJ = \cos(C)\sin(A - B)$$
$$L = GH = \sin(C)\cos(A - B)$$

Trig I.D.: $$L = \frac{1}{2}[\sin(C + (A - B)) + \sin(C - (A - B))]$$

$$K = \frac{1}{2}[\sin(C + (A - B)) - \sin(C - (A - B))]$$

The only thing that changed from Case 1 is that now K does not have A (-) minus sign.

$$\text{Output} = K + L = \sin C + (A - B)$$

This is a signal above the R.F. carrier C, except when the audio signal frequency equals the audio oscillator frequency $(A - B) = 0$. The frequency is then C.

The net result of both cases is a single sideband signal that is symmetrical about the R.F. carrier. (See Figure 10-11.)

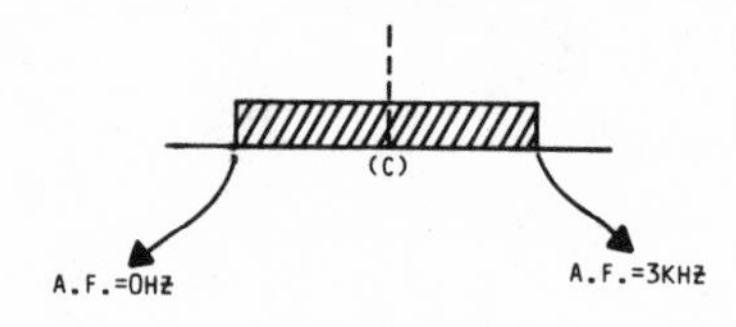

Figure 10-11

SECTION 4: I AND Q CHROMANCE SIGNALS

Introduction

The color information portion of the television video signal is transmitted as two different signals known as the I signal and the Q signal. The manner in which these signals convey color information is very involved. The following is a condensed summary of the way color information is conveyed.

Transmitted Signals

The color T.V. camera has three separate outputs, one for each color: red, blue, and green. These signals are fed to a matrix which produces the Y, I, and Q signals as shown in Figure 10-12. The Y signal contains the luminance information and is the signal detected by a BW television set. The percentages of R, B, G that make up the luminance signal are chosen to produce a black and white signal to the human eye. The minus signs in the Figure represent 180 degrees phase shift.

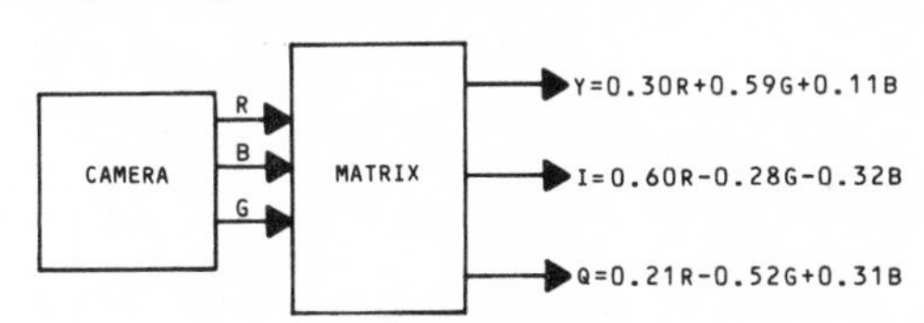

Figure 10-12

The I and Q signals are modulated onto a subcarrier as shown in Figure 10-13. The balanced modulators eliminate the subcarrier and allow only the sidebands to pass. The subcarrier is shifted 90 degrees before being modulated by the Q signal. This puts the Q signal in quadrature with respect to the I signal.

The frequencies of the horizontal oscillator, the vertical oscillator, and the color subcarrier are all related. The horizontal and vertical must be related to produce the necessary 262.5 scanning lines per field. The horizontal and color subcarrier are related so that the sidebands can be "frequency interlaced," in order to reduce beat frequencies.

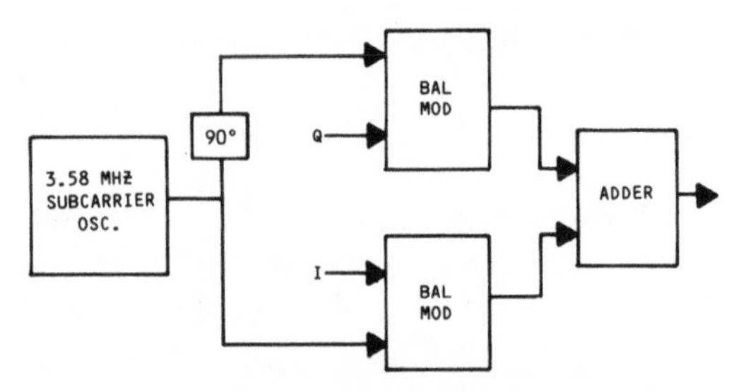

Figure 10-13. Subcarrier is 3.579545 ± 10HZ

The horizontal frequency is chosen so that an even harmonic of this frequency is equal to the 4.5 mhz intercarrier. The even harmonic that yields a horizontal frequency closest to the frequency used in monochrome broadcasting is 286. This gives:

$$F_H = \frac{4.5\text{MHZ}}{286} = 15{,}734.26\text{HZ}$$

This is because there must be 262.5 horizontal cycles per vertical cycle.

$$F_V = \frac{15734.26}{262.5} = 59.94\text{HZ}$$

The color subcarrier is chosen to be:

$$F_C = 455 \times \frac{F_H}{2} = 3.579545\text{MHZ}$$

Bandwidths

The bandwidths of the I and Q signals are shown in Figure 10-14. The I signal is allowed higher frequencies since it corresponds to the colors of orange and cyan, and these colors are resolved better by the human eye.

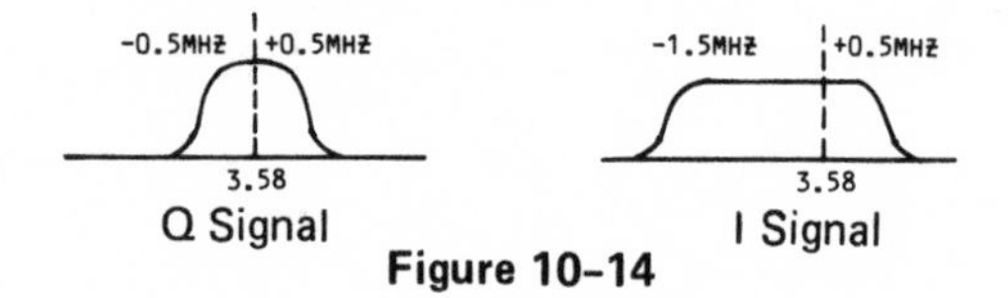

Figure 10-14

Receiver Section

The block diagram for the receiver sections is shown in Figure 10-15. It is basically a reverse of the transmit process, with the 3.58 oscillator being phase-locked to the transmitted color burst.

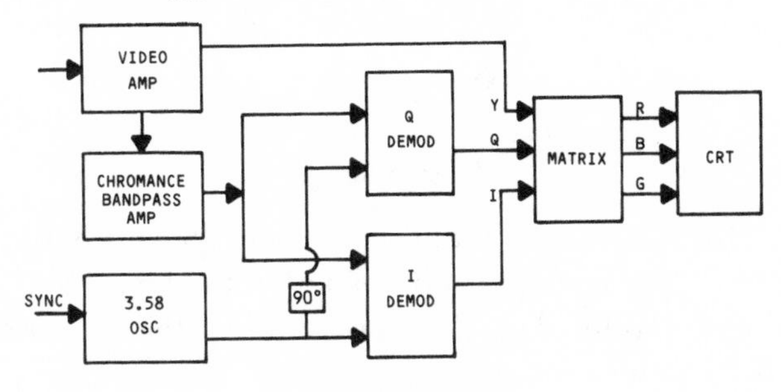

Figure 10-15

Other Equations

Signals related to Color T.V. can be written in several different ways as follows:

$$Y = 0.30R + 0.59G + 0.11B$$
$$I = 0.60R - 0.28G - 0.32B$$
$$Q = 0.21R - 0.52G + 0.31B$$
$$I = 0.74\,(R - Y) - 0.27\,(B - Y)$$
$$Q = 0.48\,(R - Y) - 0.41\,(B - Y)$$
$$R - Y = 0.70R - 0.59G - 0.11B$$
$$B - Y = -0.30R - 0.59G + 0.89B$$
$$G - Y = -0.30R + 0.41G - 0.11B$$

G - Y is normally derived from R - Y and B - Y so:

$$G - Y = -0.51\,(R - Y) - 0.19\,(B - Y)$$
$$R = Y + 0.93I + 0.63Q$$
$$B = Y - 1.11I + 1.71Q$$
$$G = Y - 0.28I - 0.64Q$$

Color Circle

The color reproduction can be represented by a color circle as shown in Figure 10-16.

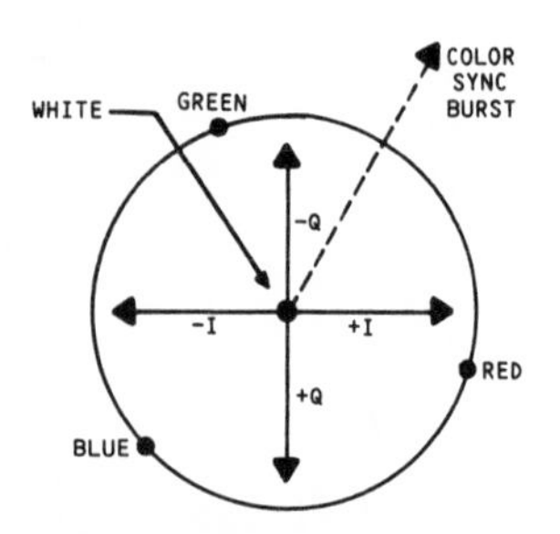

Figure 10-16

The color saturation is determined by the magnitude of the vector sum of the instantaneous values of the I and Q signals. The color hue is determined by the phase angle of this vector sum with respect to the color-sync burst. (See Figure 10-17.)

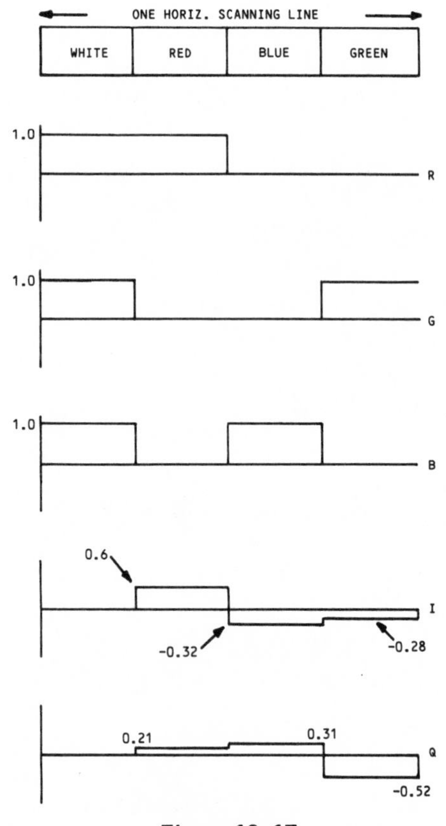

Figure 10-17

SECTION 5: NEGATIVE FEEDBACK AMPLIFIERS

The use of negative or degenerative feedback in amplifiers gives many desirable results, including reduced distortion, wider bandwidth, etc. The following material derives the equations for gain in amplifiers using negative feedback.

Voltage Feedback

The amount of feedback is determined by the values of R_1 and R_2. Let's call this value β, so that the amount of feedback voltage is βV_{out}. (See Figure 10-18.)

INPUT — V_{IN} — AMP GAIN -A — V_{OUT} — OUTPUT

GAIN = $\frac{V_{OUT}}{V_{IN}}$ = -A, THE MINUS SIGN INDICATES 180° PHASE SHIFT

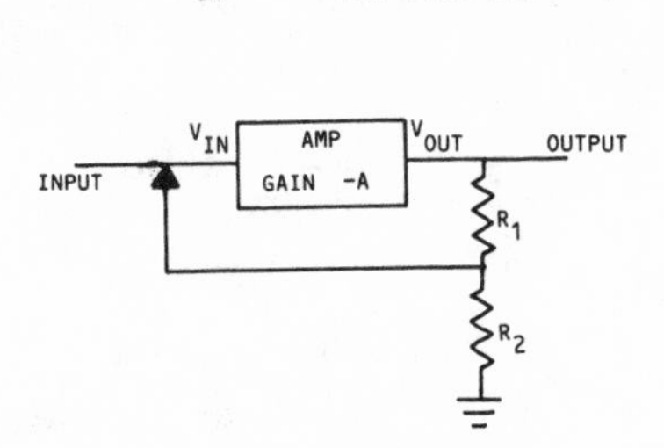

Figure 10-18

Without feedback $V_{out} = -AV_{in}$

Let: $(-A_f)$ = gain with feedback

Therefore: $V_{out} = (-A_f)(V_{in} - \beta V_{out})$

$$-A_f = \frac{V_{out}}{V_{in} - \beta V_{out}}$$

$$-A_f = \frac{-V_{out}}{V_{in} - \beta V_{out}}$$

Substitute: $V_{out} = -AV_{in}$

$$A_f = \frac{AV_{in}}{V_{in} + \beta AV_{in}}$$

$$A_f = \frac{A}{1 + \beta A}$$

Example: (See Figure 10-19.)

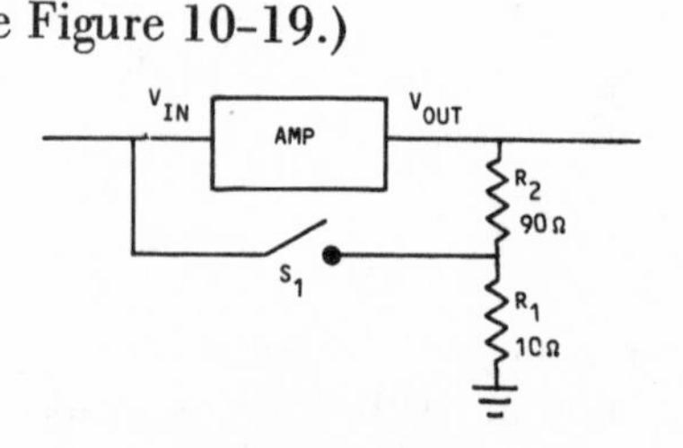

Figure 10-19

The gain of this amplifier with S_1 open is 50.

With S_1 closed the feedback ratio $(\beta) = \frac{R_1}{R_1 + R_2} = 0.1.$

Therefore: $A_f = \frac{50}{1 + 0.150} = \frac{50}{1 + 5} = 8.33$

Operational Amplifier

The equation for voltage gain of an operational amplifier is derived as shown in Figure 10-20.

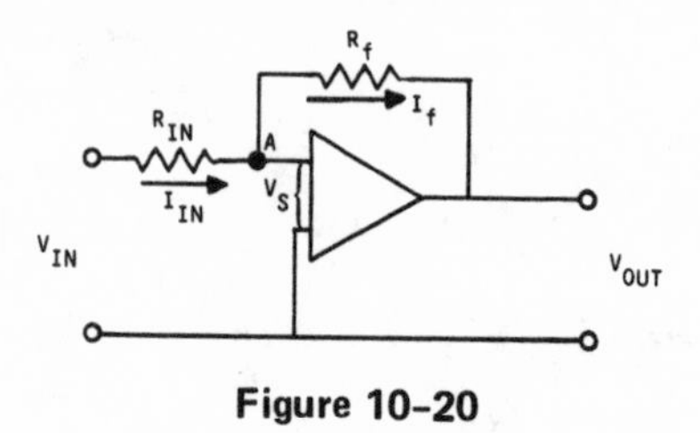

Figure 10-20

1. $V_s = \frac{V_{out}}{-A}$ — Voltage gain of the amp

2. $I_n = I_f$ — Kirchoff's law at point A

3. $I_f = \frac{V_{out} - V_s}{R_f}$ — Ohms law

4. $I_{in} = \frac{V_{in} - V_s}{R_{in}}$ Ohms law

5. $I_f = \frac{V_{in} - V_s}{R_{in}}$ Sub EQ_2 into EQ_4

6. $\frac{V_{out} - V_s}{R_f} = \frac{V_{in} - V_s}{R_{in}}$ Sub EQ_3 into EQ_5

7. $\frac{V_{out} - \frac{V_{out}}{-A}}{R_f} = \frac{V_{in} - \frac{V_{out}}{-A}}{R_{in}}$ Sub EQ_1 into EQ_6

8. Assume A is $> 10^4$.

9. Therefore: $\frac{V_{out}}{A} \longrightarrow 0$

10. $\frac{V_{out}}{R_f} = \frac{V_{in}}{R_{in}}$

11. $\frac{V_{out}}{V_{in}} = \frac{R_f}{R_{in}} = \text{Gain}$

SECTION 6: POWER CALCULATION IN HIGH LEVEL A.M. PLATE MODULATION

The First Class F.C.C. Exam contains some questions requiring knowledge of the power relationships involved in high-level plate modulation. A block diagram of this system is shown in Figure 10-21. Here are some important points to keep in mind:

1. The power that goes into the sidebands comes from the modulator (final audio amplifier).
2. Going from no modulation to 100 percent sinusoidal modulation the following things occur:
   A. Total output power goes up 50 percent.
   B. R.F. output current goes up 22.5 percent.

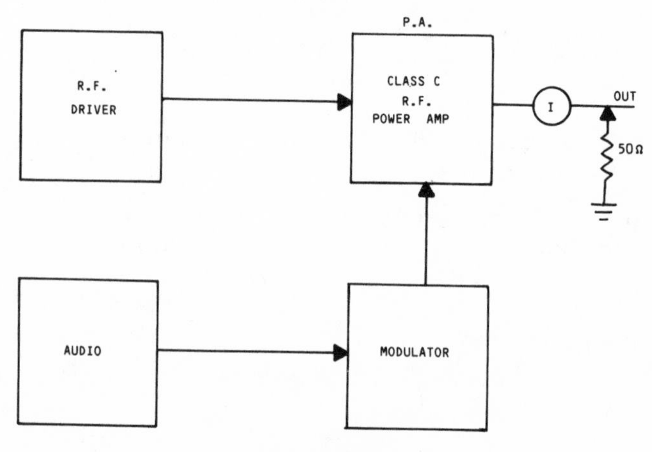

Figure 10–21

C. Total power dissipated in the R.F. amplifier increases.

D. D.C. current to the R.F. amplifier remains constant.

3. Sideband power varies as the square of the modulation percentage.
4. If the R.F. amplifier is X percent efficient, this applies to the D.C. power supply as well as audio power coming from the modulator.

Example 1:

Given: 1. R.F. output power = 100 watts.
2. R.F. amp plate voltage = 1KV.
3. R.F. amp efficiency = 70 percent.

Find: The D.C. plate current to the R.F. amp.

$$\text{D.C. input power} = \frac{\text{output power}}{\text{efficiency}}$$

$$\text{D.C. power} = \frac{100}{.7} = 142.8 \text{ watts}$$

$$142.8 \text{ watts} = \text{plate voltage} \times \text{plate current}$$

$$\text{plate current} = \frac{142.8}{1000} = 142.8 \text{ milliamps}$$

Example 2:

Given: The same conditions as Example 1.

Find: The audio output power from the modulator necessary for 100 percent modulation.

The R.F. carrier power is 100 watts. When 100 percent modulated, the total output power will be 150 watts. The extra 50 watts must come from the modulator.

The power output from the modulator must be $\frac{50}{0.7} = 71.43$ watts, since the 70 percent efficiency of the R.F. amplifier applies to audio input as well as D.C. input.

Example 3:

Given: 1. R.F. amplifier plate voltage 1000 volts, plate current 500 milliamps, efficiency 80 percent.

2. Modulator plate voltage 1000 volts, plate current 250 milliamps, efficiency 75 percent.

Find: The R.F. output current into a 50 load.

To find the R.F. current, first find the total R.F. power.

$$\text{Carrier power} = V_p \times I_p \times \text{Eff}$$
$$= 1000 \times .5 \times .8 = 400 \text{ watts}$$

$$\text{Sideband power} = \text{modulator output power} \times \text{R.F. amplifier efficiency}$$
$$= 1000 \times .25 \times .75 \times .8$$
$$= 150 \text{ watts}$$

$$\text{Total output power} = 550 \text{ watts}$$

$$P = I^2 R \text{ or } I = \sqrt{\frac{P}{R}} = \sqrt{\frac{550}{50}} = \sqrt{11} = 3.316 \text{ amps}$$

Example 4:

Given: R.F. amplifier plate voltage 1000 volts, plate current 400 milliamps, 75 percent efficiency.

Find: The total amount of power dissipated in the R.F. amplifier plate during 100 percent sinusoidal modulation.

The total power dissipated equals the D.C. power dissipated, plus the audio power dissipated.

Input power = $V_p \times I_p$ = 1000 × .4 = 400 watts

At 70 percent efficiency, 400 × .7 = 280

280 watts goes as R.F. carrier power

120 watts goes as plate dissipation.

For 100 percent modulation, the modulator output power must be 50 percent of the R.F. amplifiers D.C. input power.

Audio = 400 × .5 = 200 watts
200 × .70 = 140 watts for R.F. sideband power
60 watts dissipated
Total power dissipated in the R.F. amps plate

$$\underset{\text{D.C.}}{\underset{\downarrow}{120}} + \underset{\text{Audio}}{\underset{\downarrow}{60}} = 180 \text{ watts}$$

Example 5:

Given:
1. R.F. amp plate voltage of 2000 volts, efficiency of 75 percent.
2. R.F. output current into a 50Ω load of 5.0 amps.
3. No modulation

Find: The D.C. plate current

Total output power = $I^2 R = 5^2 \times 50$
= 25 × 50 = 1250 watts

1250 watts is 75 percent of the input power

$$\text{Input power} = \frac{1250}{.75} = 1666\frac{2}{3} \text{ watts}$$

$$I_p = \frac{1666\frac{2}{3}}{V_p} = \frac{1666\frac{2}{3}}{2000} = .8333 \text{ amps}$$

SECTION 7: TRANSISTOR BIASING

Transistors can be biased in several different ways, but there are three common methods that are usually encountered. These biasing methods can be analyzed using simple algebra to determine the D.C. voltages on the transistor terminals and sometimes the gain in voltage. By using the same algebra, it is possible to design a transistor amplifier and determine the proper resistor values for biasing it. The following examples illustrate how this is done.

One basic assumption is that silicon transistors have a forward-biased base-emitter voltage drop of 0.7 volts. Germanium transistors would have a drop of 0.3 volts, base to emitter.

A. Biasing Method 1

Equations

$V_b = 0.7$ volts

$$I_b = \frac{V_{cc} - 0.7}{R_1}$$

$I_c = \beta I_b$

$V_c = V_{cc} - I_c R_z$

Example: (Figure 10-22.) $V_{cc} = 12.0$ volts

$$I_b = \frac{12.0 - 0.7}{100{,}000} = .000113 \text{ amps} = .113\text{ma}$$

$I_c = 50 \times .000113 = 5.65$ma

$V_c = 12 - 5.65 = 6.35$ volts

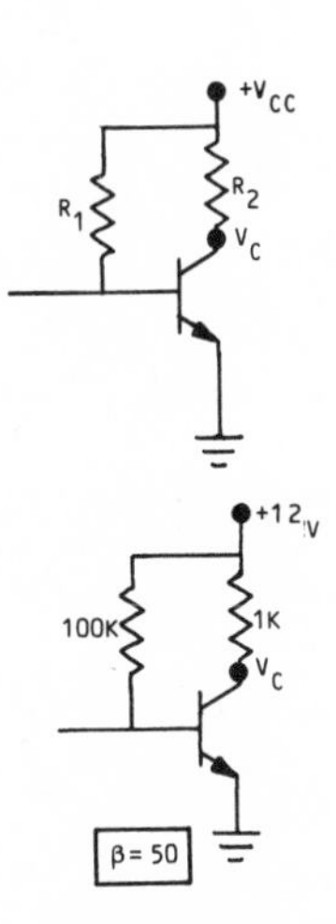

Figure 10-22

B. Biasing Method 2

Equations

$V_b = 0.7$ volts

$$I_c = \frac{V_{cc} - V_c}{R_1}$$

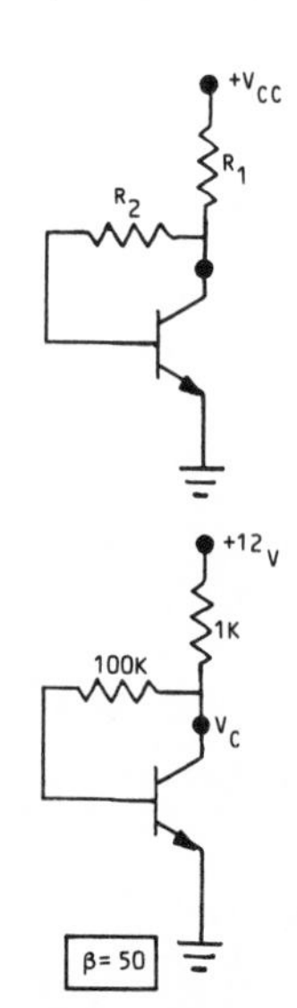

Figure 10-23

$$I_b = \frac{V_c - 0.7}{R_2}$$

$$I_c = \beta I_b$$

$$V_c = \frac{R_2 V_{cc} + 0.7\, R_1}{R_2 + R_1}$$

Example: (Figure 10-23.)

$$V_c = \frac{12 \times 10^5 + 0.7 \times 50 \times 10^3}{10^5 + 50 \times 10^3}$$

$$V_c = \frac{12.35 \times 10^5}{1.5 \times 10^5} = 8.23 \text{ volts}$$

C. Biasing Method 3

Assumptions

1. $I_c \gg I_b$ so $I_c = I_e$
2. $V_b - V_e = 0.7$ volts (silicon)

Equations: (See Figure 10-24.)

$$V_b = \frac{R_1}{R_1 + R_2} \times V_{cc}$$

$$V_e = V_b - 0.7 \text{ volts} \qquad \text{Gain} = -\frac{R_4}{R_3}$$

$$I_e = \frac{V_e}{R_3} = I_c$$

$$V_c = V_{cc} - I_c R_4$$

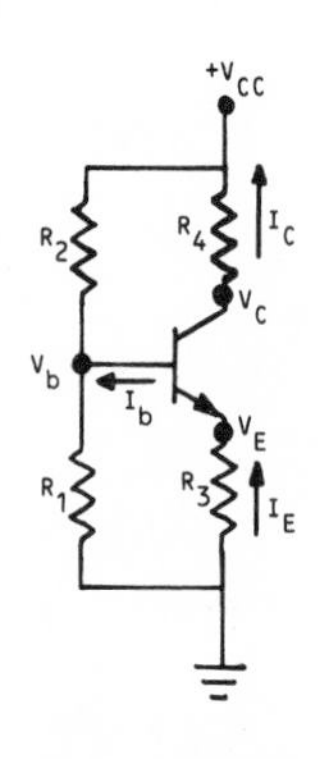

Figure 10-24

Examples: (See Figure 10-25.)

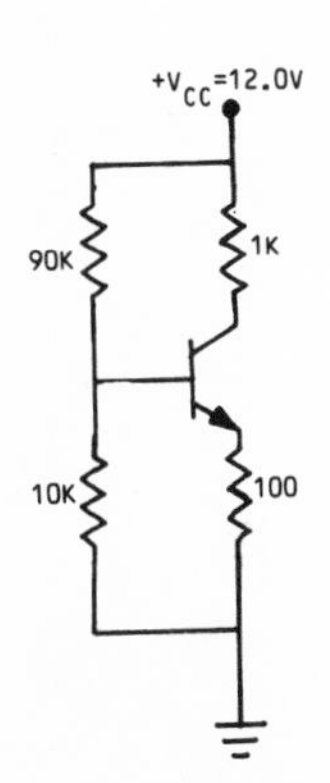

Figure 10-25

$$V_b = \frac{10}{90 + 10} \times 12 = 1.2 \text{ volts}$$

$$V_e = V_b - 0.7 = 0.5 \text{ volts}$$

$$I_e = \frac{V_e}{R} = \frac{0.5}{100} = .005 \text{ amps}$$
5 ma

$$V_c = V_{cc} - I_c R_4$$
$$= 12 - .005 \times 1000$$
$$= 7 \text{ volts}$$

Gain = 10.0

Example: Design an Amplifier with a Voltage Gain of 12.

Biasing Method 3 will be used, since it is thermally stable and the gain is independent of the transistor Beta. With a supply of +12.0 volts, the desired voltage on the collector is +6.0 volts or half the supply voltage. This allows voltage swings of ±6 volts without clipping. (See Figure 10-26.)

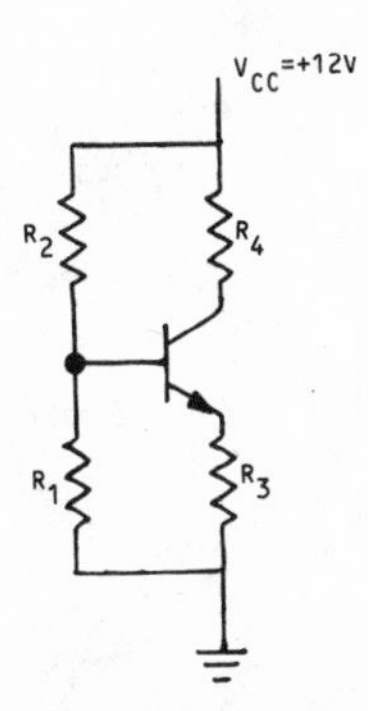

Figure 10-26

A. For a voltage gain of 12:

$$\frac{R_4}{R_3} = 12 \quad \text{or} \quad R_4 = 12R_3$$

B. Since $R_4 = \dfrac{V_{cc} - V_c}{I_c}$ and $I_c = \dfrac{V_e}{R_3}$

$$R_4 = \frac{(V_{cc} - V_c)R_3}{V_e}$$

C. We want $V_{cc} - V_c = 6.0 \text{ volts} \left(\frac{1}{2}\right)V_{cc}$

$$\text{so } R_4 = \frac{6.0R_3}{V_e}$$

and $R_4 = 12R_3$

so $\frac{6.0}{V_e}$ must equal 12

and $V_e = 0.5$ volts

D. If we let $I_c = 5MA$

$$R_4 = \frac{6.0}{.005} = 1200\Omega$$

$$R_3 = \frac{R_4}{12} = 100\Omega$$

E. Since $V_e = 0.5$ volts, $V_b = 1.2$ volts

F. $V_c \times \frac{R_1}{R_2 + R_1} = V_b$

so $12 \times \frac{R_1}{R_2 + R_1} = 1.2$

or $\frac{R_1}{R_2 + R_1} = 0.1$ (A 10-to-1 voltage divider)

G. Resistors R_1 and R_2 can have several values, each of which will work. The limits are:

- The current $I = \frac{12}{R_2 + R_1}$ must be large compared to the transistor base current.
- The current $I = \frac{12}{R_1 + R_2}$ must be kept small so that the power supply drain is small.
- $R_1 \| R_2$ determines the amplifier input impedance, or since $R_1 \ll R_2$, R_1 determines the input impedance.

Practical values for this condition are:

1. $R_1 = 10K$ $R_2 = 90K$
2. $R_1 = 5K$ $R_2 = 47K$

The first choice gives higher input impedance and lower current drain from the supply voltage.

INDEX

A

A.C. circuits, complex, 84
Admittance, 140
Algebra, 20
Algebraic operations, 26
Algebraic manipulation, 35
Algebraic substitution, 35
Alphanumeric, 218
A.M. Plate modulation, 241
Amplifier:
  efficiency, 242
  feedback, negative, 225
  operational, 240
  transistor, 225
Amplitude modulation, 229
"And" Gate, 89
Angle addition, 127
Angle:
  double, 127
  subtraction, 127
Antenna, 58, 197
Arbitrary constant, 152
ASCII, 16, 218
Associative Law, 21, 98
Audio: 232
  bypass, 232
  oscillator, 321

B

Balanced modulator, 230
Base current, 248
BCD code, 14, 15
BEL, 56, 57
Bessel functions, 179
BETA, 247
Bias voltage, 195
Binary, 6, 7, 8
  addition, 9, 10
  division, 12
  multiplication, 11
Binary (*cont'd*)
  numbers, 6
  subtraction, 10
Boolean algebra, 88-97
  rules of, 97, 98
Broadcasting, 164

C

Carrier, 180, 230
  power, 243
  sub carrier, 235
Cartesian coordinates, 40
Centimeter, 214
Chromance, 235
Coefficients, 32
Cofactors, 28, 29, 30
Color circle, 237
Commutative law, 21, 98
Complements, 17
  1's, 19
  2's, 18
  9's, 18
  10's, 17
Complex:
  A.C. circuits, 84
  conjugate, 48, 49, 50, 134
  numbers, 46, 47, 48, 49, 133
Conductance, 140
Connectives, 87
Constants, 152
  integration of, 152
  time, 160
Cosecant, 124
Cosine, 124
Cotangent, 124
Coulombs, 151, 152, 160
Cross multiplication, 28
Current nodes, 66
Current source, 62
Cyan, 236
Cycle, 130

D

Decibels, 53, 56, 57, 59, 212
Decimal, 1, 12
  addition, 9
  division, 12
  multiplication, 11
  numbers, 1
  subtraction, 10
Degenerative feedback, 239
Demorgan, 98
Denominator, 81
Derivative, 144
Determinant:
  algebra, 27
  four-by-four, 30
  high order, 27
  three-by-three, 27
  two-by-two, 27, 28
Determinants, 26, 34
Differential calculus, 144
Differential equations, 159
Diodes, 192, 194
Distorted sinewaves, 164
Distributive law, 21, 98

E

E.B.C.D.I.C. code, 219
Effective radiated power, 58
Equation, 20
Excess 3, 15
Exclusive "OR", 116
Exponents: 43, 53
  fractional, 43, 44
  negative, 44
  zero, 44
Exponential notation, 54
Euler's constant, 224

F

Factoring, 24
Farads, 147
Foot, 214
Fourier series, 169
Fourier transform, 178
Fractions, 12, 13
Free space wavelength, 199
Frequency:
  deviation, 182
  distribution, 179
  interlaced, 235
  modulation, 164
Full adder, 116
Full subtractor, 118

G

Gram, 214
Graphs, 36, 41
Gray code, 15, 113

H

Half adder, 115
Half subtractor, 117
Harmonic, 169, 171
Harmonic amplitude, 171
Henries, 158
Hertz, 130
Hexidecimal, viii
Hex-to-decimal conversion, 214
Horizontal oscillator, 235
Hue, 238
Hypercube, 110
Hypotenuse, 123

I

I-signal, 235
Imaginary numbers, 26, 45, 47, 131
Impedance, 84, 85, 140-142
  load, 200
Inch, 214
Induced voltage, 155
Infinite ladder network, 227
Infinite series, 164
Infinity, 150
Input impedance, 248
Integral calculus, 144
Integrals, ix, 148
Inverse trig functions, 126
Inverter, 89
Irrational numbers, 45

J

J-fet, 195
"J"-operator, 45, 81, 131
Joules, 147

K

Karnaugh Maps, ix, 110-115
Kilogram, 214
Kilometer, 214
Kirchoff: 34, 61
    Current law, 34, 61
    Voltage law, 34, 61

L

Line segment, 104, 107
Linear components, 37
Linear equations, 37, 38, 39
Liter, 214
Load lines, ix, 192-196
Logarithms, 53
    antilog, 54
    colog, 54
    common, 54
    natural, 54
Logic:
    positive, 99
    negative, 98, 99
Logic diagram, 87, 91
Logical:
    Addition, 88
    Multiplication, 88, 89
Low-pass filters, 232

M

Maclaurin series, 164
Main diagonal, 31
Meter, 214
Metric system, 213
Mile, 214
Minors, 28
Minuend, 10
Modulation index, 181
Monograms: viii, 184
    coil winding, 186
    impedance, 185
    Ohm's law, 187
    resonance, 189
Multiple integration, 158
Multiplicand, 11

N

N-cube, ix, 104-110
"Nand" gate, viii, 90
Negation, 89
Negative logic, 98
"Nor" gate, 90
Norton, 61
Numbers: 45
    complex, 133
    imaginary, 131
    irrational, 131
    line, 45
    rational, 131
Numerator, 81

O

Octal numbers, 3, 8
Ohm's law, 53, 60, 62
One-ohm cube, 225
Operational amplifier, 240
"Or" gate, 88
Ounce, 214

P

Parallel resistor graphs, 191
Parallel resonant circuit, 85
Parity, 119
    bit, 119
    circuit, 120
    even, 119
    odd, 119
Period, 130
Periodic waveform, 169, 170
Phase, 131
Phase shift, 235
Plate modulation, 241
Polar coordinates, 40, 41
Polar form, 47
Positional notation, 1
Positive logic, 98
Pound, 214
Product-of-sums, 95
Pythagorean theorem, 124

Q

Q-signal, 235, 236
Quadrature, 235
Quadratic equation, 23, 24
Quarts, 214

R

Radians, 126
Radicals, 43
Rational numbers, 45
Reactance, 21, 78, 134
  inductive, 21, 78, 134
  capacitive, 78, 134
Real numbers, 46, 47
Reference levels, 59
Relay, 87
Remainder, 10, 12
Resistor ladder network, 227
Resonance, 22, 84
  condition of, 23
Right triangle, 123

S

S-meter, 182
Sawtooth wave, 169
Scientific notation, 50, 51, 52
Secant, 124
Short circuit, 73
Sidebands, 229
Simultaneous equations, 39
Sine, 124
Sinewave, 128
Single sideband, 229-234
Sinusoidal modulation, 243
Slope, 144-146
Smith chart, 197
Square root, 23, 45
Square wave, 169
Substitution, 23
Subtrahend, 10
Sum-of-products, 94
Superposition, 61
Susceptance, 140
Switching function, 87, 93
Symmetry conditions, 172
Sync burst, 238

T

Tangent, 124
Tangent line, 146
Taylor series, 167
Thevenin, 61
Third method S.S.B., 231
Time constant, 160
Transistor biasing, 245
Transmission lines, 199, 200
Transmission line impedance, 184, 197
Transpose, 31
Triangle wave, 164, 169
Trig identities, 127
Truth table, 87
Tunnel diode, 194, 195

U

"Unit distance" code, 15

V

Vectors, 80, 136, 138
Vector sum, 238
Velocity factor, 199
Vertical oscillator, 235
Vertices, 105, 107
Voltage, 62, 129
  average, 129
  feedback, 239
  gain, 58
  loops, 66
  peak, 129
  peak-to-peak, 129
  RMS, 129
Volume units, 59
VSWR, 197

W

Waveform, 169-173

Y

Y-axis, 39, 41
Y-intercept, 38
Y-signal, 235-237